Social Epistemology and Technology

Collective Studies in Knowledge and Society

***Series Editor*:** James H. Collier is Associate Professor of Science and Technology in Society at Virginia Tech.

This is an interdisciplinary series published in collaboration with the Social Epistemology Review and Reply Collective. It addresses questions arising from understanding knowledge as constituted by, and constitutive of, existing, dynamic, and governable social relations.

The Future of Social Epistemology: A Collective Vision edited by James H. Collier
Social Epistemology and Technology: Toward Public Self-Awareness Regarding Technological Mediation edited by Frank Scalambrino
Socrates Tenured: The Institutions of 21st Century Philosophy, Adam Briggle and Robert Frodeman
Social Epistemology and Epistemic Agency edited by Patrick J. Reider

Social Epistemology and Technology

Toward Public Self-Awareness Regarding Technological Mediation

Edited by
Frank Scalambrino

ROWMAN & LITTLEFIELD
INTERNATIONAL
London • New York

Published by Rowman & Littlefield International Ltd
Unit A, Whitacre Mews, 26-34 Stannary Street, London SE11 4AB
www.rowmaninternational.com

Rowman & Littlefield International Ltd. is an affiliate of Rowman & Littlefield
4501 Forbes Boulevard, Suite 200, Lanham, Maryland 20706, USA
With additional offices in Boulder, New York, Toronto (Canada), and Plymouth (UK)
www.rowman.com

British Library Cataloguing in Publication Data
A catalogue record for this book is available from the British Library
ISBN: HB 978-1-7834-8532-1
PB 978-1-7834-8533-8

Library of Congress Cataloging-in-Publication Data
Social epistemology and technology : toward public self-awareness regarding technological mediation / edited by Frank Scalambrino.
pages cm. — (Collective studies in knowledge and society)
Includes bibliographical references and index.
ISBN 978-1-78348-532-1 (cloth : alk. paper) — ISBN 978-1-78348-533-8 (pbk. : alk. paper) — ISBN 978-1-78348-534-5 (electronic)
1. Social epistemology. 2. Knowledge, Sociology of. 3. Self-consciousness (Awareness) 4. Technology. I. Scalambrino, Frank, 1976– editor.
BD175.S62165 2015
303.48'3—dc23 2015032165

∞™ The paper used in this publication meets the minimum requirements of American National Standard for Information Sciences—Permanence of Paper for Printed Library Materials, ANSI/NISO Z39.48-1992.

Printed in the United States of America

Contents

Introduction

Publicizing the Social Effects of Technological Mediation

Frank Scalambrino

"Technology is not a neutral power. . . . The basic fallacy which stifles all other considerations consists in the belief that technology is to be viewed as a self-contained causal system" (Jünger 1963/1983, 277–79). This edited volume brings together scholars from across disciplines to discuss the social effects of technological mediation. The book is divided into two closely related parts. The first part includes chapters related to the normative social dimensions effected by technological mediation of knowledge. The second part includes chapters related to the changing conceptions of humans and humanity influenced by technological mediation.

As a volume in the *Collective Studies in Knowledge and Society* series, this book directly participates in three of the five activities targeted for the series. They are: (I) Promoting philosophy as a vital, necessary public activity; (II) Analyzing the normative social dimensions of pursuing and organizing knowledge; and (III) Exploring changing conceptions of humans and humanity. Whereas both the content and the very existence of this book participate in the first of the targeted activities, the parts of the book are divided, respectively, across the other two activities. For the sake of heuristic convenience, we may say that the first part of the book places emphasis on the relation between technological mediation and epistemology, understood as first and foremost *social epistemology*, and the second part of the book places emphasis on the social effects of technological mediation understood in terms of ontology. Yet, of course, one of the virtues of social epistemology is that it does not attempt to exclude ontology from its studies regarding knowledge and society.

OVERVIEW OF THE THEMES

The title, and especially the subtitle, of this book speaks of a number of themes regarding which some introductory comments may be helpful. First, we take the theme of social epistemology to be grounded in the following "fundamental question":

> How should the pursuit of knowledge be organized, given that under normal circumstances knowledge is pursued by many human beings, each working on a more or less well-defined body of knowledge and each equipped with roughly the same imperfect cognitive capacities, albeit with varying degrees of access to one another's activities? (Fuller 1991, 3)

Additionally, we understand "social epistemology" as an interdisciplinary field of study developing through collective efforts and avenues such as the *Social Epistemology Review and Reply Collective* (http://social-epistemology.com/), *Social Epistemology* (Taylor & Francis), and *Episteme* (Cambridge). The collective and interdisciplinary nature of social epistemology further highlights the efficacy of *publicizing* studies regarding knowledge and society "toward a revival of the normative in the sociology of knowledge" (cf. Fuller 1991, 265) as opposed to a logic or *technique* of justification (cf. Guston 1993). To acknowledge that how the pursuit of knowledge is organized matters is not so much to democratize truth as it is to highlight that decisions made regarding labor, material resources, and time contribute largely to determining what truths are discovered, thereby conditioning the norms governing society and identifying the individuals who people it.

In other words, the normativity associated with social epistemology refers to dialectic and discovery more than the norms of demonstration. Which is to say that *we* collective members of society, through our organizations and the organizing of our pursuit of knowledge, dynamically and reciprocally participate in the construction of society. In this way, policy decisions normatively shape contingent existential and historical possibilities for societies to the extent of determining constraints on lived experience, which also means, to an extent, determining constraints on individual freedom. On the one hand, here "dynamic" and "reciprocal" point to the problematic phrase "public self-awareness" in this book's subtitle. On the other hand, of course, technology plays the role of a nonneutral power in mediating both one's lived experience and the knowledge production and dissemination, which sculpt the landscape of everyday existence in a society. Hence, examining technological mediation, in this context, contributes to "analyzing normative social dimensions."

Further, the following chapters, as is consistent with traditional philosophy of technology, focus on the qualitative, rather than the quantitative, dimensions of the social effects of technological mediation. In other words, following Plato, Aristotle, Martin Heidegger, and Jacques Ellul, for example, "technology" is to be distinguished from "technological devices." Moreover, the quantitative/qualitative distinction is part of the Western philosophical tradition systematized by Immanuel Kant (cf. 1998). Hence, when specific technological devices and applications are discussed in the following chapters, such discussions are usually in the service of evidencing prevailing interests and historical cultural tendencies. That is to say, just as being able to operate a motor vehicle involves prerequisite knowledge, so too does operating a smartphone. Beyond the quantitative questions regarding the construction or number of motor vehicles or smartphones, we are interested in the more qualitative questions regarding, for example, the constitutive and influencing effects of the presence of motor vehicles and smartphones (cf. Hilkins and Cherwitz 2011; cf. Browning 1990).

"Technological mediation," then, refers to the use of technology to mediate various relations to reality. This includes human to human, human to self, and human to natural and social forces, for example, legal and economic. What is more, as will be discussed further below, technological mediation may be characterized in terms of qualitative degrees. What this means is that at a depth of technological mediation the very identity of that to which one is technologically related changes in terms of the technological mediation. When such a depth of mediation has been reached, the relation may be referred to as "cybernetic." Below, we will discuss the difference between theoretical and applied cybernetics. Lastly, that the spectrum along which a relational phase change from non-cybernetic to cybernetic occurs is qualitative may be illustrated in a twofold way. First, the quantity of technology involved is not so much the determining factor as the manner in which the technology involved reveals the identity of that to which it is related. Second, though of the reality in question we say that it is the same both before and after some technologically mediated relation to it, we also say that were it not for the technology mediating the relation, the cybernetic dimension of the reality in question would not be present to us.

Lastly, consider the complicated and problematic phrase "public self-awareness." Simply put, how is "public" to be understood? The term is intended here to convey the same meaning it does in the locutions "public defender," "public notary," "public service," and "public domain." Such terms are less problematic than the related and likewise operable uses of "public" in "public trust" and "public interest" (cf. Fellows 2014; cf. Serrano and Bonilla 2007). On the one hand, the integrity supposed to be associated with the role of an official appointment to serve the public as an impartial witness, for example, a public notary and a public defender, is involved in the concepts of public "trust" or "interest." On the other hand, just as *we* collective members of society dynamically and reciprocally participate in the construction of society, the idea of a "public interest" risks slipping from the revival of the normative to a technique of justification to answer questions such as "Who determines what is in the public's interest?" or "What betrays the public's trust?" (cf. Syn 2014; cf. Radder 2008). Therefore, consider the following clarification regarding "public self-awareness."

Similar to Plato's famous comparison of types of city with types of soul from the *Republic*, "public" may be initially understood as it relates concretely to property (1997c). For example, a section of land may be designated "public property" if it is neither privately owned nor restricted from public use (cf. Sax 1970). Similarly, "domain" may refer concretely to a "territory over which dominion is exercised," or more abstractly to "a sphere of knowledge, influence, or activity" (Merriam Webster 2015). A more *abstract* rendering of public property, then, may refer to a dimension that theoretically would be inextricably nonexclusive; yet, the question whether such a "public domain" exists remains open (cf. Encabo and Martin 2007). After all, "public" may be inextricably tied to a particular hegemony of thought (cf. Gramsci 1971; cf. Scalambrino 1998). Put another way, in the words of Steve Fuller and James Collier, "One way to look at the fate of the public sphere is that the 'forums' that used to focus the life of a *polis* are now commercially licensed to the media for purposes of mass consumption" (2004, 235).

Yet, at the same time, "conceptual support" for a legal notion of "public trust" may be illuminating. According to one legal theorist, "The approach with the greatest historical support holds that certain interests are so intrinsically important to every citizen that their free availability tends to mark the society as one of citizens rather than serfs" (Sax 1970, 484). This distinction between "citizens" and "serfs" maps with the traditional distinction generally in Western philosophy, and particularly in the philosophy of technology, between "persons" and "subjects." Moreover, just as these distinctions map respectively with what would be a movement from revived normativity to a logic of justification, it is meaningful to inquire regarding awareness in the reciprocal process of norm construction.

Further, a concept of the public may be in the service of resisting or counteracting influence from various "private interest" groups, thereby further enabling openness and integrity in policy determination and decision making (cf. Sax 1970). For example, according to the United Nations Environment Program (UNEP), "Public awareness refers to the important role that community enthusiasm and knowledge has in building sustainable societies" (UNEP 2015). Now, in regard to an individual person, "self-awareness" may be understood as "private" or "public." Whereas private self-awareness refers to an awareness not publicly accessible, public self-awareness refers to a socially dependent awareness. Hence, in regard to knowledge and society, "public self-awareness" may refer not only to a publicly accessible dimension of social knowledge to which we may contribute but also to a domain in the service of freedom to which collective members may relate to increase in public self-awareness at the individual level (cf. Carson 1962; cf. Zea 1987; cf. Agassi 1989; cf. Serchuk 1989). Such a domain of freedom does not exclude normative shaping, rather it resists the inauthenticity (cf. Heidegger 1962, 234) of machination and objectification, sustaining personhood against influences with the potential to diminish its integrity, such as technological mediation (cf. Brey 2008).

OVERVIEW OF THE CHAPTERS

The first part of this volume includes chapters related to the normative social dimensions effected by technological mediation of knowledge. This part opens with a chapter by Steve Fuller titled "The Place of Value in a World of Information: Prolegomena to Any Marx 2.0" in which he "offers a genealogy of the concept of information" discussing the manner in which it is embedded in "platform capitalism." According to Fuller, "Platform capitalism presents a subtler version of Marxian exploitation that does not rely on a clear sense of employment or wage." Just as technological mediation has effected a shift to information processing as a primary mode of production, it may be understood in Marxist terms of commodification, that is, "a product that arose as a means to facilitate socially useful ends becomes an end in itself, bending all the activities that it supports in its own image." After discussing the history of the "value of information" toward the relativization that conditions its current technologically mediated value in, and for, platform capitalism, the chapter concludes with a discussion of changes for which a public might advocate toward self-awareness and to lessen potential exploitation in light of the growing commodification of cybernetic "state-istics," that is, the value of use information for governments, such as a principle of "*coveillance*."

The next chapter "Technological Systems and Genuine Public Interests" is by Hans Radder. In response to what he describes as "the strongly increased commodification of academic science," he discusses the necessary conditions for substantiating the claims of several authors that such commodification represents a genuine public interest. He discusses the nature and role of large technological systems that constitute what he calls "a society's sociomaterial infrastructure." As if "too big to fail," such systems have an epistemological and ontological social efficacy that extends across time. Radder discusses how such systems "form social resources," which necessarily "represent certain interests." Yet, he critically notes, these interests neither coincide nor may be reduced to the interests of individual members of a society.

Hence, just as not all of these "supra-individual" interests are genuine public interests, such supra-individual interests critically address their use in the justification of increased commodification of academic science.

In "The End of Trust in the Age of Big Data?," Daniel Jo Brunson uses concepts from the social epistemology regarding risk management to explore the changing parameters of trust in the face of increasing technological mediation. Brunson discusses the "information asymmetry" given that in "the age of Big Data we are monitored as never before." This information asymmetry results from the relation between citizens of a society and the technological mediation that surveils them. Further, this asymmetry shifts empowerment away from citizens. Brunson discusses the manner in which this shift makes citizens vulnerable and the conflict of interests involved in determining the extent to which "trust" may characterize the process.

In the fourth chapter, "Filter Bubbles and the Public Use of Reason: Applying Epistemology to the Newsfeed," Jamie Carlin Watson discusses the relation between technological mediation as it relates to the fundamental question of social epistemology and censorship. That is, technological mediation in the form of "filter bubbles" or "filtering algorithms" have an implicit claim to public interest and trust in the form of efficiently catering to the informational needs and knowledge requirements of citizens by organizing search results and, thereby, one's pursuit of knowledge. Though the process of information filtering may claim to be in the service of efficiency and individual well-being, on the one hand, it may amount to a process of censorship. On the other hand, it may amount to a process of "persuasion profiling" predicated on bias or private interests. Watson argues that this process should be submitted to public scrutiny.

The next chapter, "The Internet and Existentialism: Kierkegaardian and Hegelian Insights," is by Patrick J. Reider. He takes Hubert Dreyfus' use of Søren Kierkegaard to critique the internet regarding anonymity, commitment, and trust, as a point of departure to develop existentialist themes that reveal the internet as a threat to personhood. Reider (re)introduces G. W. F. Hegel for the sake of critiquing Kierkegaard's "failure to appreciate the role of social mediation." Acknowledging that in the "internet age" the relation between "knowing, as a form of social mediation, and effective agency is obscured," he suggests that Hegel's concept of personhood may, on the one hand, ameliorate some existential concerns regarding the internet and the internet age. On the other hand, he underscores the need for social epistemology, in the face of such technological mediation, to determine meaningful action.

In "Existential Privacy and the Technological Situation of Boundary Regulation," Elize de Mul argues against those who, in light of the increasing technological mediation of social life, advocate for the elimination of privacy. Noting that proponents of social media, such as current Facebook CEO Mark

Zuckerberg, make the mistake of thinking that privacy *can* be eliminated, de Mul, following the work of philosopher and biologist Helmuth Plessner, argues that "privacy is an inherent ontological characteristic of human being." Emphasizing Plessner's, over Heidegger's, account of communal dwelling, that is, the *Mitwelt*, she shows how Plessner's idea of "excentric positionality" not only "makes public self-awareness possible and inevitable" but also reveals privacy as an existential aspect of human being. As such, the project of privacy elimination through technological mediation is really an issue of public boundary regulation.

In the seventh chapter, "Critical Media: Media Archeology as Critical Theory," Stephen M. Bourque explores the intersection between media archeology and critical theory to develop a sense of "critical media." According to Bourque, "Media archeology may be understood as the application of critical theory to the effects of technological mediation, both historically and socially." Advocating for critical media, this chapter discusses its methodology and capacity to engage the interdisciplinary nature of the material of social epistemology. That is, the scope of critical media includes the diverse multiplicity of discourses, ways of practicing, and methods of inquiry involving both the humanities and sciences. Hence, Bourque argues that critical media may be a means through which public self-awareness regarding technological mediation may emerge.

The next chapter, "Speculative Ethics and Anticipatory Governance of Emerging Technology: A Case for 'Un-disciplined' Philosophy of Technology," is by William Davis. Regarding the technologies that mediate (social) reality, he notes that "their social, economic, and political impacts resist early identification and tend to be realized only after technologies have permeated societies and cultures." Hence, if philosophers of technology and social epistemology are to engage in efficacious deliberations, then the contributions need to be made in the early phases of the development of new technologies. In this way, Davis advocates for speculative ethics and anticipatory governance, and he holds that the context for such deliberations should be public. This is the thinking behind "un-disciplined" philosophy of technology. Technological mediation has made the public convening of groups for the purpose of fostering "education, motivation, and participatory governance projects" easier to organize. Including such underrepresented voices allows for the individuals through whom the social effects of technological mediation manifest to contribute to the processes with which their lives and the lives of future generations will be, perhaps unavoidably, involved.

The final chapter of this section is titled "What Control? Life at the Limits of Power Expression." This chapter discusses how normative social dimensions have been effected by technological mediation of knowledge at the level of the individual. Though freedom entails some degree of control, existential freedom is the very condition for the possibility of being under control, and this chapter characterizes the substitution of freedom subjected to control for existential freedom as a central feature of what social epistemologists and critical philosophy of technology theorists have called "the society of control." Contributing to the illustration of cybernetics as the condition for the possibility of such substitution, through a brief discussion of critical moments in the history of philosophy, cybernetics appears along a spectrum of technological mediation, as if effecting a gestalt-like paradigm shift in regard to self-knowledge mediating action with existential freedom as a point of departure. Invoking Plato, Aristotle, Ernst Jünger, Martin Heidegger, Herbert Marcuse, and Gilles Deleuze to criticize cybernetics, especially its application in psychology, this chapter concludes by advocating for self-awareness in regard to existential freedom, toward reorienting an inauthentic relation to power expression.

The second part of this volume includes chapters related to the changing conceptions of humans and humanity influenced by technological mediation. This part opens with a chapter by Charles Bambach, "Heidegger on *The Question Concerning Technology* and *Gelassenheit*," in which he explicates Martin Heidegger's "interpretation of modern technology." As the everyday term "human resources" may suggest in our time, Heidegger's interpretation of modern technology warns that the technological mediation of human experience may put humans in the service of a technologically-mediated life. That is to say, in the undertow of instrumentalization, relating to reality as an instrument or means to an end, and machination, relating to reality as a cybernetic machine process, we relate to our own being as such. We orient our lives in terms as if our existence were a kind of stockpile. In light of such technological mediation, we run the risk of being merely resources for employment, that is, means to an end. This chapter ends with a discussion of what may be characterized as Heidegger's articulation of an antidote to the pervasive technological mediation of our time, that is, poetic dwelling and the Gelassenheit of contemplative meditative thinking.

The next chapter "How Learning to Read and Write Shapes Humanity: A Technosomatic Perspective on Digitization" is by Joris Vlieghe. He argues beyond mere instrumentalization that technological mediation is more than a means. Taking as a point of departure the idea that it is only through education that one is able to accomplish personal formation, Vlieghe argues that with the digitization of education "a new form of human subjectivity might emerge." The substantial contribution that technological mediation makes in education, then, is such that a distinction may be made between traditional and digital literacy. Vlieghe contrasts these two types of literacy discussing the production of letters and the creation of meaning. To the extent that digital literacy subjects humans to a different type of formation, then we may expect to see a different type of subjectivity emerging in human history, especially in terms of how these subjects relate to creative acts and meaning making.

In "Labor and Technology: Kant, Marx, and the Critique of Instrumental Reason" Arthur Kok focuses on the relation between work, that is, human labor, and human well-being. Whereas labor allows humans to satisfy material needs, achieve a level of comfort, and to realize freedom, the increasing instrumentalization and mechanization of labor disrupts such beneficial outcomes, according to Karl Marx. After explaining what he takes to be "the core of Marx's critique," of the instrumentalization and mechanization of labor, Kok invokes Immanuel Kant's concept of labor to examine whether a dimension of the realization of freedom through labor may be salvaged in the wake of increasing technological mediation. For Kant, labor is necessary for self-awareness, and though inducing such labor requires coercion, the coercion as Kant characterizes it must be explicitly *not* economic or instrumental.

In the thirteenth chapter, "The Biopolitics of the Female: Constituting Gendered Subjects through Technology," Danielle Guizzo analyzes the role that technological mediation plays in reinforcing what she refers to as "biopolitical forms of control" such as regulation of the female body and gender. For instance, Guizzo points to birth as something to be controlled through the technology of chemical mediation, that is, the birth control pill, indicating a socially constituted knowledge of female behaviors as requiring social regulation and administration. On the one hand, the biopolitics of the female, then, provides an example of how technological mediation influences the historically changing conceptions of being human. On the other hand, the biopolitics of the female provides an example of how socially constituted understandings of the female gender, including "the discourses and truth systems which shape and regulate the female body," organize the pursuit of social knowledge. Technological mediation not only sustains what may be characterized as a cybernetic rendering of the female but also actually *shapes* the physical and lived experience of female bodies.

In the next chapter, "Phenomenology of Radiology: Intentional Analysis in the Constitution of Diagnostic Judgment" Mindaugas Briedis brings together two fields of research and practice, that is, phenomenology and radiology, to discuss how technological mediation "may enhance an epistemological relation to absent Others and allow for an empathic connection with Others who are merely technologically present." Briedis' phenomenology is decidedly Husserlian, and, as such, he examines the role of "intentional analysis" in forming diagnostic judgments through radiological images. Briedis explicitly discusses "the role imaging plays in the constitution of the intended object, the role of embodiment, imagination and empathy." Though the radiological requirement of technological mediation in forming a diagnostic judgment speaks to the constitution of an object otherwise inaccessible to human experience, he emphasizes the role that such technological mediation plays in a process of human caregiving.

In "Absent to Those Present: The Conflict between Connectivity and Communion," Chad Engelland discusses the difference between connectivity and communion. He then examines degrees of personal presence, ranging from "in person" to increasingly technologically mediated degrees of presence. He argues that social technology "makes us present to those absent," while it "makes us absent to those present." Maintaining that "bodily presence is needed for intimacy and friendship," Engelland holds that the "interpersonal presence afforded by social technology is a presence purchased at a price," which is "the absence of the body." Here, the distinction between connectivity and communion brings forth the sense in which bodily communion is missing despite the connectivity of social technology. Self-awareness founded on the connectivity of technological mediation over the communion of bodily presence may constitute a changing conception of humans and humanity in the field of social epistemology or beyond.

In the sixteenth chapter, "Recognizing the Face and Facial Recognition" Levi Checketts "examines the implications of facial recognition software, embedded in augmented reality glasses, to examine the question of social responsibility." Whereas the designers of such technological mediation consider such mediated encounters to be superior to nonmediated encounters, Checketts criticizes the claim through the work of Jean-Paul Sartre, Gabriel Marcel, and Emmanuel Levinas. His path moves from facial recognition software by degrees through the notions of "the look" (Sartre) and "the encounter" (Marcel) toward what he takes to be the most appropriate understanding of "the face" (Levinas). Though facial recognition software may instantly provide instrumental social knowledge of the other person encountered, such mediation, Checketts concludes, may work more as a barrier than as a catalyst in recognizing the face of the other.

In the next chapter, "Situated Mediation and Technological Reflexivity: Smartphones, Extended Memory, and Limits of Cognitive Enhancement," Chris Drain and Richard Charles Strong focus on technological mediation in relation to groups of people. They argue that "technologies, and the practices they facilitate in their mediating roles, can co-constitute communities." These technological mediations change how "we act as social, cognitive, agents." Alterations or enhancements to human capacities such as knowing and remembering are bound up with technological practices illustrated by the 'smartphone.' Though the smartphone may be understood as "extending" one's mind, that is, mental capacities, Drain and Strong consider how such an enhancement contributes to changing conceptions of humans and humanity when a situation is mediated by multiple individuals with such "enhancements." Given such conditions, they predict changes to self and social identity resulting from such changing interpersonal and collective modes of cognition. From a systems-based perspective, taking the assemblage or group as more primary than the individuals that comprise it, extended and enhanced cognition would distribute differently across the group than would nontechnologically mediated cognition.

The final chapter of this section is titled "The Vanishing Subject: Becoming Who You Cybernetically Are." This chapter discusses changing conceptions of humans and humanity influenced by technological mediation, specifically in terms of relational ontology and applied cybernetics. Articulated with Aristotle's vocabulary of practical action and Immanuel Kant's vocabulary of "regulative" and "constitutive" ideas, this chapter explores how the unfreedom associated with applied cybernetics can permeate to the level of constitutive changes for humans in multiple ways. Because existential freedom pertains to one's potentiality, the result of cybernetically mediating one's relation to one's existential freedom amounts to embracing a version of one's self determined by the technological cybernetic mediation, rather than one's ownmost potential, and such a relational alteration does have constitutive effects. In fact, such effects take hold through the instantiation of mediated habits and mediated knowledge, thereby ultimately redirecting the freedom of one's will toward a virtual reality. This chapter invokes, as examples, procreation and the constitution of "lived experience." Hence, in the attempt to become who one cybernetically is, one subjects oneself to cybernetic objectification such that this subjectivity vanishes into knowledge of a virtual objectification and a habit structure constituted through cybernetically mediated regulative ideas, that is, impersonating another in the effort to impersonate oneself.

Part I

NORMATIVE DIMENSIONS OF TECHNOLOGICAL MEDIATION AND PUBLIC SELF-AWARENESS

Chapter 1

The Place of Value in a World of Information

Prolegomena to Any Marx 2.0

Steve Fuller

This chapter offers a genealogy of the concept of information, which today is very much embedded in "platform capitalism," whereby keystrokes and clicks operationalize "the difference that makes a difference," which by the mid-twentieth century had become the default position of a radically relativized understanding of the "value of information" (aka meaning). The first half of the chapter deals with how information became so relativized, which goes back to the general collapse in the distinction between short- and long-term assessments of value in the wake of Kant's demystification of theodicy. Marx, fortified by Hegel, tried to keep this distinction alive, especially via the concept of exploitation, whereby whatever short-term benefit workers receive from their employers is supposedly undermined by a longer-term disadvantage (aka surplus value). Platform capitalism presents a subtler version of Marxian exploitation that does not rely on a clear sense of employment or wage. Moreover, Capitalism 2.0 replaces the physical space that was expropriated in Capitalism 1.0 with a possibility space that is equally ripe for expropriation. In this context, the concept of "digital prosumer" is a significant corrective measure to enable people to regain control of their data.

1. THE PLACE OF VALUE IN A WORLD OF INFORMATION

The title of this section echoes the title of the 1934–1935 William James Lectures delivered at Harvard by the *Gestalt* psychologist Wolfgang Koehler, "The Place of Value in a World of Facts." Koehler's use of the word "value" would have been understood by Aristotle, for whom value was to be found in an organism's state of equilibrium with its environment, which he understood as an optimal integration of the organism's many functions. This was the "wisdom of the body" which the Harvard physiologist Walter Cannon had dubbed "homeostasis" a few years prior to Koehler's lectures. For his part, Koehler was specifically focused on what it would mean for the brain to be in such a state of dynamic equilibrium with its environment. Implied in this entire line of thought is an equivalence of "value" and "health," such that the healthy organism is performing as designed—or put in less explicitly genetic terms, the organism is being the best version of itself that it can.

The Christian cosmos complicates this Aristotelian take on things. Put more provocatively, there is no natural affinity between Aristotle and Christianity, which helps to explain the energy that the Roman Catholic Church has had to expend to keep the two aligned, despite the many fronts of dissent (aka heresy). After all, for the Christian, the true value of any creature can only be understood in ultimate, or "absolute," terms—that is, once one understands the divine plan within which the creature's existence fits, which in turn resolves all the messiness of life as lived. It follows that "homeostasis" is a temporary, not a final, state of being. In this respect, Christianity introduces a radically instrumentalist conception of value, as we are all means to God's ends. However, humans remain unique because they are more than merely divine instruments. They are "ends in themselves," as Kant famously put it, because we have been created *in imago dei*. But that simply means we enjoy (or endure!) a divided ontological status: Like the deity from whom we descend, we create things that aspire to ultimate value; yet at the same time we remain the most dutiful, albeit often unwitting, vehicles for realizing God's intentions. The Christian aims to reconcile the two perspectives, with Jesus serving as the paradigm case (Fuller 2011, ch. 2).

In practice, this means zooming in and out between a micro and macro view of things, which suggests that the value of an action changes over time—before, during, and after it happens—as it comes to be seen in light of other actions. Thus, in the moment of *theosis*, when the "transfigured" Jesus realizes that he is the Son of God, he glimpses his own relatively casual demise on the cross followed by a more substantial redemption. This sensibility was captured more abstractly in the late seventeenth century as "theodicy," the branch of theology devoted to "justifying the ways of God to men." Although Voltaire's *Candide* ridiculed this vision for reducing all suffering to a failure to perceive a deferred good, less than a century later it was secularized in Hegel's philosophy of history as the "cunning of reason," which then made its way to Marx's dialectical materialism. But of equal importance is theodicy's legacy to classical political economy, which secularizes and democratizes humanity's divided ontological status—now, in terms of two opposing sources of value. On the one hand, *the labor theory of value* is a secular and democratized version of the idea that human action is naturally endowed with godlike creativity; on the other hand, *the utility theory of value* is a secular and democratized version of the idea that the ultimate significance of human action is only known in the long run by the benefit that others derive from it.

An ideal market may be seen as capturing, through the price mechanism, the state of play in the dialectic between these two conceptions of economic value. Thus, our potentially boundless creativity is disciplined by the specific wants of others that need to be served, while at the same time those wants become more refined as consumers are forced to decide from among the plethora of products at their disposal. In this way, naïveté and excess come to be focused into efficiently taken decisions. As a result, the economy's set-point is continually shifted as people become more sophisticated in terms of what they produce and consume (Fuller 2006b). In this context, Condorcet's conception of markets as a vehicle of overall human progress proves more illuminating than that of his contemporary, Adam Smith (Rothschild 2001). Condorcet's use of *tâtonnement* ("groping") for the stabilization of prices—what Léon Walras characterized as "equilibrium"—was also used by various Lamarck-inspired thinkers to characterize the microstructure of evolution (Fuller 2015, 162ff). Thus, the meaning of each individual's life amounted to the management of one's personal supply and demand curve. One recent philosopher who has drawn on this insight to ground a general metaphysics of value has been Robert Nozick (1981, ch. 5). Unfortunately his efforts have gone largely ignored, even though they could provide the normative basis for a "deep liberalism" (Fuller 2015, 126ff).

The neat story I have been telling of the secularization of theodicy into the dialectics of economic value has yet to take account of Kant's influence, which was to radicalize the distance between the two poles of the dialectic, rendering each inscrutable to the other. Once Kant's "transcendental" turn reduced our centrality to the cosmos from necessary truth to wishful thinking (in the existence of God), the value attached to human labor became attenuated to simply one's ownership of the decisions taken to act in an uncertain world, while "utility" evaporated into "fate," as the consequences of our actions came to be seen as fundamentally beyond our control. Thus, whatever efficacy—let alone legacy—that might be ascribed to the human condition on the basis of art or technology was ontologically undercut. From a materialistic standpoint, Kant seemed to reduce human history to a series of ephemeral states, and the price mechanism itself reduced to the fleeting outputs of a stock exchange ticker tape. In this way, the path from Kant (via Kierkegaard's critique of Hegel) to existentialism was paved.

The academic side of this Kantian dismemberment of value is enshrined as the familiar fact-value distinction from the philosophy of social sciences. It is often presented as a secular dogma that serves to stave off the dreaded "naturalistic fallacy," which threatens to derive an "ought" from an "is." Thus, in a remarkable inversion of the original theology, the sociologist Max Weber described the mixing of facts and values—as, say, economists do when they presume the inherent goodness of productivity increases—as the "original sin" of social scientific inquiry (Proctor 1991, 89). *Contra* Weber, one might have thought that like Adam's expulsion from Eden, the traumatic moment was precisely Kant's own expulsion of values from the world of fact. In any case, it was against this Kantian context that starting in the mid-nineteenth century, under the influence of Hermann Lotze, it became common to speak of "value" in the plural as "values." It is easy to overlook the significance of this move, which may be regarded retrospectively as the first step toward the "postmodern condition" (Lyotard 1979/1984). The grammatical innovation signaled the radical privatization of value, with the burden of proof shifted to those who would make claims for "common value," perhaps even including "public good" (Fuller 2015, 109ff).

However, it would be a mistake to regard the fact-value distinction as requiring the *subjectivization* of values, in the sense of immunizing them from scientific treatment. On the contrary, Lotze himself had been a student of Gustav Fechner, the founder of experimental psychophysics, a discipline dedicated to determining the change in a physical stimulus needed to make a "just noticeable difference" in an organism's sensory response (Heidelberger 2004). It would be more correct to say that the neo-Kantian turn initiated by Lotze involved simply the *relativization* of values, which may be understood just as objectively as any other empirical phenomenon—namely, as the products of specific social conditions or personal histories or even, as B. F. Skinner might put it, "schedules of reinforcement." Thus, two individuals or cultures may hold radically different values, but both can be understood empirically and comparatively. Whether these individuals or cultures can live amicably in the same society without one or both undergoing radical existential change is another matter entirely, of course. Indeed, this was the context in which the problem of "cross-cultural rationality" was raised in the twentieth century, typically under the shadow of imperialism (e.g., Wilson 1970; Hollis and Lukes 1982).

Perhaps the most lasting impact of this value relativization has been felt through the account of information provided by the Shannon-Weaver model of communication, which locates the value of a transmitted signal in what the anthropologist Gregory Bateson famously dubbed "the difference that make a difference" to a receiver's decision-making capacity. While such differences can be empirically specified (i.e., the new input that decides a matter for a particular receiver), they cannot be universalized or perhaps even generalized very widely across individuals. However, that has not stopped the concept itself from becoming generalized—and popularized. When the Shannon-Weaver model was first proposed in the 1940s, it was seen as providing an operational definition of "meaning" (Gleick 2011). More recently, in a world increasingly saturated with data in search of inferential levers, the phrase "tipping point" has gained currency to capture what statisticians call "criticality," the moment when a change in degree becomes a change in kind (Gladwell 2000).

A decade before the popularization of "tipping point," and still thinking in terms of psychophysics, I coined the term "axioaetiotics" for the study of how value judgments change in response to changes in one's sense of the causal structure of history (Fuller 1993, 169ff). In particular, the more the value of an action or event rises, the more necessary it appears to have been to the realization of a valued outcome. (Imagine the value of cause X rising as the prior probability of a highly valued effect Y decreases.) Conversely, claims that an outcome was "overdetermined" lower the value of its actual causes because other historically available factors would have sufficed to produce something comparable. Anyone familiar with the passing fashions in historiography, several of which exist simultaneously at a given time, knows that intuitions about the value-cause nexus are quite fluid, yet they often determine how we evaluate present-day claims to legitimacy.

The set of historiographical intuitions that have most concerned me turn on the Scientific Revolution's historic dependency on certain factors uniquely operating in seventeenth-century Europe. Put the other way around: Would China, by all standards the world's superpower until 1800, have eventually had its own scientific revolution—even without Western interference? The persistent return of a negative verdict by historians of science suggests that the nature of science is indeed connected with something distinctly Western. I have pursued this point at length elsewhere (esp. Fuller 2015, ch. 6). However, one implication worth flagging here is that if the significance of the Scientific Revolution has been as great for world history as is normally supposed (e.g., Herbert Butterfield thought it second only to the rise of Christianity), then it becomes at least in principle justifiable to impose science on cultures that would otherwise never have come to it (aka imperialism). The Scientific Revolution would seem to be the difference that has made all the difference—at least in terms of providing the framework for making sense of the subsequent course of world history, especially if told as the story of our self-transcendence of our species limitations.

In the second half of this chapter, I explore the normative questions that arise when the sort of value associated with an information-based standard of meaningfulness—"the difference that makes the difference"—becomes the foundational principle of political economy. This is arguably the world in which we find ourselves today. It is a version of "Capitalism 2.0" that calls out for a "Marx 2.0" as its deep but sympathetic critic who points to a normatively acceptable "Humanity 2.0" (Fuller 2011). To be sure, the global financial crisis of 2008 has opened up a nostalgia industry in Marxist critiques of capitalism worldwide, ranging from the ultra-serious David Harvey to the ultra-comic Slavoj Zizek. But what is really needed is a reinvention of Karl Marx's positive vision suited to our times. In contrast to Thomas Piketty, the French economist who was dubbed by *Time* magazine on May 8, 2014 "Marx 2.0," Marx 1.0 did more than bemoan rising wealth inequalities. He identified capitalism as housing the potential to radically transform the social order for the better. "Socialism" in Marx's view would succeed by *completing* capitalism rather than merely immunizing people from its worst effects, à la today's version of "social democracy," let alone eliminating capitalism altogether (Fuller and Lipinska 2014, ch. 3).

2. PLATFORM CAPITALISM'S CHALLENGE TO MARX 2.0

In the early days of the World Wide Web, a brave book entitled *Cyber-Marx* used Marx's own writings, especially the *Grundrisse*—the prototype for *Das Kapital*—to infer what Marx would make of the emerging centrality of information and communication technologies to the capitalist mode of production (Dyer-Witheford 1999). What is perhaps most striking about this book to readers nearly two decades later is its adherence to the Capitalism 1.0 narrative. *Cyber-Marx* continues to make much of Marx's original (correct) prediction that capital's domination of labor would in the future come less from managing the time of workers than from replacing them with machines that can do the requisite work more efficiently than the humans could. However, the book fails to consider that new opportunities for extracting surplus value might emerge from the spontaneous activities of people spending an increasing portion of their lives—for both work and leisure—in front of a computer screen. Whereas Marx projected a proletarian revolution on the basis of the difference between the employed and the unemployed becoming more pronounced, we in fact have come to inhabit a world in which the conceptual difference between employment and unemployment is increasingly blurred, yet surplus value continues to be extracted apace—indeed, arguably at unprecedented levels. Welcome to Capitalism 2.0!

A good way to see Marx's blindspot is to consider the average person in an average job, which she performs simply to make ends meet in order to engage in activities that give meaning to her life, most of which generate recordable and surveyable data traces, typically in the form of keystrokes and clicks resulting from searches and eventuating in purchases and other "revealed preferences," as economists say. Indeed, such a person may end up generating more value in activities that she would understand as leisure than at work. This is especially true, if one considers the multiplier effect of her data, once recombined with the data of others, circulating through various algorithms that propel the search engines used by cognitively, economically, and politically enhanced agents. In short, Marx had not anticipated that the raw material of capitalism would shift from natural resources to information.

In many respects, this shift amounts to talking about the same things under different descriptions, as in the case of living things being understood nowadays in terms of their "genetic code." Indeed, human thought itself is increasingly commodified in ways that Marx never imagined. This process began in the early twentieth century with political polling and commercial advertising, both of which are now seen as versions of "market research." Over the past century, these operations have become more concentrated in their focus, less obtrusive in their operation—but no less disruptive in their consequences. The easy trade in intelligence between Silicon Valley and the US National Security Agency makes the point. Nowadays Google is allowed to operate freely in the marketplace of ideas because it produces a wealth of information from spontaneously generated online responses that can then be mined for intelligence purposes, subject to the ad hoc constraints that the state places on Google's operations. I shall have more to say about the character of these operations below.

To be sure, Marx had noticed something similar about the mutually favorable relations between the political and business classes in the nineteenth century. Back then, the state granted a business a license to expand its markets, which resulted in an increased flow of capital, which in turn benefited the state in the form of taxes. The cost, of course, was that most of the workers who were attracted to these novel arrangements—typically some form of factory work—were not protected from their worst consequences. While Marxists believed that the state was either too weak or insufficiently motivated to support the capitalist system through its inevitable periods of volatility, Bismarck realized that more was to be gained in the long term by preempting this prospect through a tax-based universal "safety net" that is now regarded as the origin of the welfare state.

This upscaling of the state's responsibility for public well-being opened the door to the mass surveillance activities that generate the data that are routinely consulted to provide a just distribution scheme of costs and benefits across the entire society. Once these data were constituted as "statistics"—that is, etymologically speaking, "state-istics"—which in previous centuries had been compiled in actuaries, hospitals, and parishes, they were consolidated and supplemented by an intensified national census that spawned survey-based research (Porter 1995, Part I). Together these streams have provided steady sources of data for the social sciences—and the clearest confirmation that the aspiration to universal knowledge and universal governance of the human condition go hand-in-hand. Generally speaking, this information has been extracted, either by explicit consent or more implicitly as a condition for receiving a service. To be sure, there have been violations at both the consent and the service ends of this process, most problematically in the case of eugenics projects designed to improve both knowledge of human genetics and delivery of welfare services. Today's regimes of research regulation associated with "institutional research boards" at universities began as direct responses to these excesses (Weindling 2004).

The shift to information processing as the primary mode of production has generated a more comprehensive version of such "panopticism." Nevertheless, it falls squarely within a Marxist scenario of commodification: namely, *a product that arose as a means to facilitate socially useful ends becomes an end in itself, bending all the activities that it supports in its own image*. Marx developed his original paradigm case, money, in the context of explaining how payment started as a mark of respect for the a piece of remarkable writing—Milton's *Paradise Lost* was Marx's example—but then became a business in which authors published at fixed rates on demand (Robinson 2015). So too is the world of *platform capitalism*, whereby such information-communication platforms as Facebook and Google have gone beyond enhancing socioeconomic life; rather, they are supplanting the public sphere, such that the default settings of personhood are shifting so that "private" has come to be seen as a retreat—if not a hiding—from one's "public" online identity (Morozov 2015b). In effect, the meaning of "virtual reality" is shifting, rhetorically speaking, from oxymoron to pleonasm. Thus, we communicate, search, and buy in the "smart environments" designed by the platform capitalists that channel our behavior into data streams that are ready-made for processing by various state and business clients. Indeed, the biggest problem facing platform capitalists is that the surfeit of information they are capable of extracting—so-called "Big Data"—is increasingly delegated to algorithms that they themselves barely understand (Mayer-Schönberger and Cukier 2013).

Like the first phase of panoptical politics, today's platform capitalism is seen as a bonanza for social science research, now based on the unobtrusive character of data collection occurring during—rather than merely before or after—the provision of a public service (Watts 2014). This specific form of commodification relates to what the economist Michael Perelman (1991) originally called *metapublic goods*, namely, the data generated from how a product is used, which in turn can be used to improve the product and/or to monitor its users (Fuller 2002, ch. 1.6). Nowadays we would speak of metapublic goods as "open source"—not to be confused with "public" in the sense of "public goods." While metapublic goods are generated in public view, not all members of the public are equally well placed to turn them to their advantage—nor should one expect the state to repair this "market failure" by making the goods universally available. Thus, you may discover a bug in a Microsoft program, but at most you help others fix the problem and perhaps get some compensation from Microsoft, while Microsoft itself now has a pretext for introducing a successor software line that all of you will purchase. Indeed, in recognition of this fact, Apple describes its users as a "community," no doubt in part to soften the blow of exploitation implied in this state of affairs.

The general lesson here is that the value of metapublic goods is a function of the power asymmetries that emerge from the data contributed to a platform. Thus, Facebook can reap greater benefits from your keystrokes and clicks than you can because it can compile your data into various bespoke data bundles that serve the interests of clients in marketing, intelligence gathering, and, indeed, empirical social research. The last case has proved especially interesting and controversial, pointing to the subtlety with which platform capitalists expropriate user information. The norms governing informed consent in research are effectively circumvented once users agree as a condition of service to accept service upgrades that are partly based on user data (Morozov 2015a). In that case, what is the difference between testing the platform for potential bugs by, say, manipulating the rate of the feedback that users receive and testing users' emotional dispositions by, say, slowing the rate for positive feedback and boosting the rate for negative feedback spontaneously generated by platform traffic? There is no difference, according to the Cornell University researchers who conducted just such a mass experiment on Facebook (Segelken and Shackford 2014).

In our relatively libertarian times, this expanded panopticism tends to be cast as a violation of *privacy*. However, truer to Marx's spirit is to see it as a new form of *exploitation*. Put bluntly, "privacy invasion" has become the new "worker exploitation," which in turn reflects the inherently insecure boundaries of the self. We are thus back to John Locke's original definition of "person" as a forensic concept whose exact identity only becomes clear once there is a need to assign responsibility for action. After all, when Facebook allowed the mass manipulation of its users for research purposes, it was exploiting a legally permissible consequence of its contracts with those users. Indeed, the lawyer Lawrence Lessig (2005) has presciently argued that as social life migrates from public space to cyberspace (reflecting, as we have suggested, a shift in the locus of personal meaning), the legal standing of personhood changes from self-ownership to a form of license or rent, in which your own freedom is tied to that of the platform provider, who is the legal owner. To be sure, your interests are often aligned with those of your platform provider—after all, you both suffer from program hackers and internet viruses. However, your autonomy is compromised, à la medieval Europe, where the fates of "lord" (platform provider) and "serf" (user) were indeed bound together, but in a state of hierarchical dependency that threatens to turn information capitalism into "information feudalism" (Drahos and Braithwaite 2002).

But how exactly could we reach this paradoxical state of affairs, whereby a putative extension of capitalism brings us back to feudalism? The answer takes us back to the original Enclosure Acts that starting in the early seventeenth century converted common land in England to private property. This is normally seen as the irreversible legal step from feudalism to capitalism. Yet, it is often overlooked that privatization began as a policy for waste management because the situation that the ecologist Garrett Hardin (1968) canonized more than three centuries later as "the tragedy of the commons" had already become the norm, namely, that putatively free land was losing its value by virtue of its users—both rich and poor—not taking personal responsibility for its maintenance (McCloskey 1975). Thus, private property was an invitation to take a standing problem with no inherently prescribed solution and convert it into a legally recognized solution, one potentially with added benefits for the owner. This became the template for addressing not only new frontiers that initially present themselves as risky prospects, but more generally *the future* as such, which typically enters the horizon of the present via unintended consequences of actions taken in the past.

However, the virtual space of the internet differs from physical space in the greater heterogeneity of its potential realizations. Put more prosaically, the internet provides a platform for many more different things to happen than a piece of farmland. To be sure, the same plot of earth may support a variety of crops and livestock, each of which serve human needs and the general ecology quite differently. But this is nothing like the range of activities potentially supported by a platform like Google, Facebook, or Amazon. Even as these platforms scale up their own energy requirements, the efficiency savings that are made by civil society activities migrating to them (e.g., communicating, searching, shopping, even voting) means that in the future previously discrete features of social life will be more easily combined, or "hybridized," by virtue of their co-location. This is one relatively mundane consequence of the "converging technologies" meta-narrative that has been dominant in science policy circles since the beginning of the current millennium, which in its most extreme and perhaps popular form is supposed to be driving humanity toward "singularity" (Fuller 2011, ch. 3).

In short, whereas Marx's original gaze was set on the Capitalists 1.0 who took advantage of enclosed land to build factories and cities that radically reorganized the landscape of Britain, the gaze of Marx 2.0 needs to be set on the Capitalists 2.0 who are taking advantage of cyber-enclosure in our time to launch platforms that effectively shape not physical space but *possibility space*: a second-order version of intellectual property, if you will. However, it would be a mistake to see platform capitalism as marking the start of the information mode of production. On the contrary, it continues a development that began with the *de facto* commodification of biological science heralded as "bioprospecting": the idea that the code-like genome might be treated as a "new frontier," open for business to those willing to invest in searching for advantageous molecular combinations (Gilbert 1991; cf. Fuller and Lipinska 2014, ch. 4). Moreover, this frontier spirit is quickly spreading into neuroscience, as the brain comes to be seen as the ultimate platform, that is, the most concentrated physical space for exploring possibility space (Dunagan 2014). Thus, while the advanced technologies used to scan and probe the brain may only marginally improve our understanding of, say, established neural disorders or even our response to consumer products, they nevertheless create a much clearer cartography that opens up opportunities, if not incentives, to experiment with an organ whose full range of functions and capabilities remains largely unknown.

In conclusion, it is worth observing that while highly critical of Capitalism 1.0, Marx did not rush to a blanket condemnation. Rather he envisaged how the industrial mode of production might be turned to everyone's advantage. To be sure, it required a shift in power relations but not in the mode of production itself. The same applies for Capitalism 2.0. The power asymmetries are already in place, though the modes of exploitation are somewhat subtler. In particular, the concept of metapublic goods suggests that whatever personal benefits users get by contributing their data to a platform, these are swamped by the surplus value that platform owners can generate simply by virtue of mass data compilation and recombination. Once again, the genetic code set the precedent, this time with the IBM-*National Geographic*-sponsored Genographic Project, which since 2005 has encouraged three-fourths of a million people in over 140 countries to supply personal genetic data—both DNA samples and family histories—in service of charting the ultimate migration history of humanity (Fuller 2006a, ch. 8). More recently, the Beijing Genomics Institute (BGI), a public-private partnership, has started a "Cognitive Research Lab" that aims to sequence the genomes of thousands of high-IQ people worldwide, with an eye to finding significant patterns that might be transferable to the general population—a source of alarm for some who see it as a thinly veiled Chinese eugenics program (Miller 2013).

But of course, even if the official aims of these projects are not fully realized, there are many other commercially viable biomedical uses to which the same data may be put. Thus, exploitation in Capitalism 2.0 may amount to a failure to be included as a shareholder in any of these projects. Suppose you are a platform capitalist, and I am one of your regular users. The problem is less that you know something about me that I do not want you to know, but that this knowledge might disadvantage me, especially if I am unable to derive benefit from the fact myself. Not only do we rarely realize that our data is being mined, but we also have no natural way of acquiring access to the information gathered about us. *Wired* magazine founder Kevin Kelly (2014) has helpfully proposed a principle of *coveillance*, which would enable you to know what others do with what they know about you. To be sure, the legal innovation required to make such a principle operative is formidable. It amounts to no less than deconstructing the "property" side of "intellectual property." One salutary move in this direction is for people to be enfranchised as "digital prosumers," capable of using the platform-based data that we generate as capital available for investing in a personalized "futures market," to which others may also contribute (Louvieris et al. 2015).[1]

NOTE

1. An early version of this chapter was delivered to the Centre for Social Informatics at Edinburgh Napier University in May 2015. Special thanks to Alistair Duff and Hazel Hall for their hospitality and feedback.

Chapter 2

Technological Systems and Genuine Public Interests

Hans Radder

In response to the strongly increased commodification of academic science, several authors argue for the continued significance of public-interest science. To substantiate such arguments one needs, first, a convincing conception of what is a public interest; second, one has to explain how science can support public interests. In this chapter, I address the first requirement. My basic claim is that the best argument for the existence of genuine public interests derives from the nature and role of large technological systems.

I discuss the problem of the nature of public interests in two steps. Section 1 demonstrates the existence and importance of supra-individual interests by refuting the individualist doctrine that all interests are, or can be reduced to, individual interests. A key role in this argument is played by an analysis of the interests constituted by large technological systems. Next, section 2 argues for the view that supra-individual interests should be understood as being public interests. The role of democracy proves to be crucial in answering the important questions of who constitutes "the public," what may count as public interests, and how particular public interests should be advanced.

1. FROM INDIVIDUAL TO SUPRA-INDIVIDUAL INTERESTS

1.1 Authors who argue for the significance of science in the public interest (e.g., Krimsky 2003, ch. 11; Radder 2010, 18–20; Moriarty 2011) assume that there are genuine public interests. In contrast, quite a few social and moral theorists subscribe to the view that, basically, there are only individual interests, that is, the interests of actual, individual people. Proponents of this view can be found among classical liberalists, neoliberalists, libertarians, rational-choice theorists, and the like.[1] My focus in this chapter is on "individualist liberalists," that is, on those individualists who ground their individualism on a liberalist account of individual freedom and autonomy. To be sure, the individualist liberalists acknowledge that, in actual social practices, the interests of individual people are "aggregated." Yet, such aggregated interests are claimed to be reducible to some kind of summed-up individual interests. The position is aptly summarized in this quotation from Richard Box:

> In the aggregative perspective, the public is a collection of separate, individual thinking units holding predetermined preferences that may be altered through advertising, marketing, or propaganda. These individuals may join (interest) groups to increase the political probabilities of achieving their goals, but the result remains a function of aggregated individual preferences. To the extent the public interest is relevant to this perspective, it consists of whatever the majority of the people want at a given time, a utilitarian calculation of measurable and additive individual "positions." (Box 2007, 587)

Hence, genuine public interests—by which I mean a specific kind of supra-individual interests that cannot be reduced to a mere aggregate of the interests of a *fixed* group of *contemporary* individuals—are claimed not to exist. Therefore, a first requirement of a defense of public interests is the refutation of the individualist claim.

The basic structure of the argument of the individualist liberalists is straightforward. The individual person is taken to be the fundamental ontological unit (in the relevant areas of politics, economics, ethics, and culture). Therefore, society is no more than the mere sum of its individual members. The interests of individuals derive from their basic status as free and autonomous agents. As such, they possess a diversity of inherited and acquired properties and preferences. Taken together, these properties and preferences define their private interests.

Thus, the basic interest of individuals is to further their freedom and autonomy. Of course, there is more to the individualist account of human interests than just this bare claim. One way of developing the claim is through qualifying it in several ways (see, e.g., Hayek 2006). *First*, individualist liberalists admit that there are two basic causes that unavoidably limit the freedom of a particular individual: nature and the freedom of other individuals. For a start, nobody can be free to do things that go against the constraints of nature, including human nature. The second fundamental limitation is that the freedom of one individual may be at the expense of the freedom of another individual. For this reason, most individualist liberalists admit that certain conditions (to be provided by the state) are needed to safeguard the freedom of individuals. However, a consistent individualism implies that the impact of the state on the life of individuals should be limited to the realization of those conditions that enable an optimal freedom for all separate individuals. Hence, these conditions do not constitute genuine supra-individual interests. Both theoretically and practically, they are subordinated to the individual or aggregated interests of particular people. *Secondly*, individualist liberalists usually assume that freedom and autonomy can only be assigned to adult and healthy human agents, and not to "infants, idiots or the insane." *Finally*, they emphasize that the notion of a free and autonomous agent should be seen as an idealization. Even if it does not apply to all individuals or not always to any individual, it still is *approximately* true that (adult and healthy) human beings are free and autonomous.

1.2 I will now provide a specific critique of individualism based on the characteristics of large technological systems. First, consider this example. Suppose that a certain individual—let me call him Mr. Freeman—has been terribly annoyed by the role of large multinational banks in creating the financial crisis of 2008. Moreover, he is even more fed up by the fact that neither these financial institutions themselves nor the national governments or international regulatory agencies have taken fundamental steps toward abolishing, or at least curtailing, the current, free-floating financial system that has become largely independent of the real economy of goods and services. In addition, Mr. Freeman realizes that it is, in part, his own savings and pension money that is being abused for the speculative endeavors of unscrupulous hedge funds and for paying the bonuses of the ever-greedy CEOs of the big financial organizations. Being in such a state of mind, he concludes that he doesn't want to have anything to do anymore with this kind of financial system. Since he feels that he is, or should have the right to be, a free and autonomous human agent, he decides to give up his bank accounts, to conduct no further transactions with banks, and from now on, to handle all his monetary dealings in cash.

Alas, poor Mr. Freeman will soon discover the disastrous consequences of what seems to be a mundane decision concerning a rather limited aspect of the way he prefers to lead his life. It is quite probable that he will lose his job, because almost all employers insist on paying his salary through a bank account. What is more, currently all providers of basic services (including electricity, gas, water, internet, and telephone) do not allow cash payments anymore, since their technological systems have made this kind of transaction impossible. And so on and so on. Clearly, under present circumstances it is virtually impossible to lead a normal life without a bank account. Therefore, there is a fundamental flaw in the reasoning followed by our brave and principled Mr. Freeman. Apparently, he is not at all free and autonomous, even in the case of his preference for cash payment, a case that, prima facie, looks like a relatively minor, nonexistential matter.

This argument shows that Mr. Freeman should be advised not to take his name too literally. It demonstrates the strongly constraining impact of the financial system on the options that are open to real individuals in our current society. Moreover, the argument can be generalized from the financial system to all kinds of *large technological systems* (in short, LTSs; see Hughes 1983, 1987). In addition to the systems for financial transactions, they include transportation (e.g., roads, railways, airfare), energy (e.g., gas, electricity), water management (e.g., drinking water, irrigation, flooding protection), communication (e.g., post, telephone, internet), health care (e.g., primary care, hospitals, insurance regulations), and so on.

So far, I have intentionally emphasized the *constraining* dimension of LTSs. The reason is that this aspect often tends to be ignored or played down. But of course such systems also possess an *enabling* dimension. Thus, LTSs have a dual nature. On the one hand, because of their *generality* these systems enable a variety of uses for a diversity of purposes. Moreover, their benefits are not limited to present users with their current purposes, because the relevant systems may also be employed by future users who will possibly have quite distinct purposes. In this respect, LTSs possess a significant open-endedness, which can be exploited by future generations in order to cope with novel, and unforeseen, challenges. On the other hand, such systems also have a certain *specificity*, which implies that they are never all-purpose systems, in the sense of being compatible with any individual preferences or sociopolitical aims. A large-scale infrastructural system for car transport embodies different values and enables different choices from a dense public railway network. Furthermore, an exclusive focus on these two systems, especially in densely populated urban regions, excludes or impedes the option of bicycle transport for the lack of safe and efficient bicycle paths.

LTSs can be, and have been, studied from various disciplinary perspectives. For the purpose of this chapter, there is no need to review the literature in detail. What I will be using are the following three general conclusions. First, technological systems are *not natural* systems but consist of networks (i.e., systems of devices and their connections) designed and made by human beings. Second, like all technologies, these systems are *socio*-technological. As a brief reflection on the above examples will show, the stable and reproducible functioning of LTSs necessitates the extensive realization of both material and social (including economic, legal, cultural, and moral) arrangements and relations. Third, once extended technological systems have been realized on a large scale, they possess a certain *inertia*. As Arne Kaijser (who uses the term "infrasystems") states,

> These material frameworks have a high degree of inertia over time. The infrasystems and the built environment in which we live today are the result of

> decisions and efforts made decades and even centuries ago. Likewise, decisions and efforts we make to build and rebuild structures will shape the material world for future generations. (Kaijser 2003, 153; see also Radder 1996, 143–47)

Therefore, such systems cannot be changed overnight and, for almost all of their time, individual members of society will be stuck with the systems available to them.[2] Put differently, at any particular moment in time these systems constitute an unavoidable material and social *inheritance* with which all individual people have to cope. It is important, however, to understand that the inertia of LTSs is not absolute, in the sense of constituting an unchangeable natural barrier. These systems are mostly pre-given and cannot be changed by the will of any individual, but they are nevertheless contingent. As Hughes (1987, 80) phrases it, LTSs possess a considerable momentum, but they are not autonomous. Specific systems *emerge* as the long-term results from the actions of a large number of actors. As such, this implies the possibility of realizing alternative systems. Yet, this realization is a long-lasting process, which entails that a part of the involved individuals may never see them realized in their lifetime.

1.3 To see the implications of this analysis for the issue of individual or public interests we first need a closer look at the notion of an interest. In ordinary life, we say things like "*x* has or possesses an interest in *y*." Here, *y* is a matter of importance, or concern, to *x*. This matter of concern is mostly of a somewhat general nature. Alternatively, we also say that "*y* constitutes an interest for *x*." While *possessing* an interest is often taken to be a characteristic of an actual individual *x*, the *constituted* interests *y* are best interpreted in terms of relations. That is to say, we should not reify them into "intrinsic" interests by leaving out the "for *x*" (even if, as we will see below, we need to be much more inclusive than individualist liberalists on the question of who may qualify as such an *x*).

As I argued in section 1.2, LTSs at once enable and constrain particular types of action of individual people. As such, they *form social resources and hence constitute certain interests*.[3] These constituted interests, however, are supra-individual, in this specific sense: they do not coincide with and cannot be reduced to (a fixed aggregate of) the possessed interests of (the majority of) the *current* members of society, for two important reasons, which both derive from the inertia of the relevant LTSs.

First, the existence of the constituted interests depends, at least in part, on the past decisions and actions of individuals who are not alive anymore. Since the individualist-liberalist account of human interests does not include these individuals and since the preferences of present-day members of society cannot be assumed to coincide with those of the past, the latter are not represented in the individualist calculation of measurable and additive interests. Second, currently existing LTSs will continue to enable and constrain the actions of future individuals. However, the preferences of future individuals cannot be measured and hence are not taken into account in the individualist aggregation procedures. The overall conclusion is that the interests constituted by LTSs (which may be called "solidified interests") are genuinely supra-individual. Put differently, the constituency of constituted interests includes two groups of "nonexisting" individuals: the past individuals who do not have interests anymore and the future individuals whose interests are as yet indeterminate.

One might suggest, however, that adding the qualifications of the basic individualist-liberalist doctrine, reviewed in section 1.1, could rescue it without having to assume the existence of supra-individual interests. Let us see whether this is the case. Consider first the role of nature and of other individuals. My point here is that the constraints imposed by LTSs are of a separate kind, distinct from the two kinds of limitation taken into account by individualist liberalists. It is obvious that these constraints cannot be reduced to nature, since they derive from *socio*-technological systems. Furthermore, as we have just seen, such constraints are not exclusively due to the interference of present-day individuals either. The second qualification of the basic individualist-liberalist doctrine limits freedom and autonomy to the healthy and adult members of society. Clearly, this limitation does not apply to people like our thoughtful Mr. Freeman, who definitely is as healthy and adult as one might be. The final qualification states that claims concerning freedom and autonomy are approximations or idealizations, which mostly (but not always) apply to actual individuals. It will be clear, though, that this amendment does not help to refute the "argument from LTSs" either. After all, it is certainly not the case that Mr. Freeman will be able to do without any banking transactions most of the time. Similarly, the claim that most, though not all, of Mr. Freeman's fellows will be able to cope freely and autonomously without such transactions lacks any plausibility.

In sum, the arguments set out so far reveal a fundamental problem of the individualist-liberalist doctrine. In this doctrine, the constraints that human agents encounter in trying to realize their freedom and autonomy are limited to either the power of nature or the forces that derive from the self-interested actions of specific other human agents. However, such a conception either overlooks the strong constraints imposed by the existence of large-scale and long-term LTSs or it implicitly, but wrongly, presupposes that individual people can somehow overcome the socio-material forces exerted by these systems. Thus, a fairly "liberal" individualist like F. A. Hayek does take account of the constraints of both nature and nurture. He includes as nature "geographic conditions such as climate or landscape" and as nurture "the family, inheritance and education" and even "cultural and moral traditions"

(Hayek 2006, 78). What he does not mention, however, let alone acknowledge, is the crucial impact of large technological systems.[4]

1.4 The doctrine that social reality is made up of free and autonomous individuals, each with their own individual interests, can be, and has been, criticized from different perspectives. Some of these criticisms simply *posit* the existence of supra-individual or public entities and interests, and proceed from there to develop an alternative view of the relation between individuals and society. Although such critiques may be valuable, they are also vulnerable. To show this, consider a *formally* similar argument concerning the existence of supra-individual entities that can be found in Thomas Mann's novel *Doctor Faustus*. In a discussion among a group of students, set in 1904, one of the characters provides the following argument for the supra-individual nature of the state.

> The legitimacy of the State resided, he said, in its elevation, its sovereignty, which thus existed independent of the valuations of individuals, because it—very much in contrast to the shufflings of the Contrat Social—was there *before* the individual. The supra-individual associations had, that is, just as much original existence as the individual human beings. (Mann 1968/1949, 119)

As to its content, however, there are two important differences with the argument from LTSs. First, the latter argument does not refer to "the state" and hence it applies more broadly, also to societies without a nation-state. Secondly, and more importantly, the empirical claim of the existence and impact of LTSs does not itself depend on strong philosophical assumptions. Hence, it is less vulnerable to counterarguments, such as the one put forward by one of the other characters in Mann's novel:

> One must look very hard at the new groupings [such as the nation-state, H.R.], which today, when liberalism is dying off, are everywhere being presented, to see whether they have genuine substance, and whether the thing creating the bond is itself something real or perhaps only the product of, let us say, structural romanticism, which creates for itself ideological connexions in a nominalistic not to say fictionalistic way. (Mann 1968/1949, 120)

In contrast, inert LTSs are robust socio-material arrangements, rather than structural ideas that could be criticized from a nominalist perspective as being mere fictions.

The conclusion of this section is that there are supra-individual interests in the following specific sense. In order to realize their goals, the individual members of a society depend on the presence of particular LTSs. These systems enable and constrain particular types of action of the individual members of society. As such, they form social resources and constitute interests. However, these solidified interests cannot be reduced to (an aggregate of) the possessed interests of the current (adult and healthy) members of society.

2. FROM SUPRA-INDIVIDUAL TO PUBLIC INTERESTS

2.1 In theory, a supra-individual interest may be a privatized, a communal, or a public interest. We are dealing with a privatized interest if all decisions concerning the design, realization, and maintenance of the LTSs providing the relevant social resources are taken to be the exclusive responsibility of particular private enterprises, for instance, for-profit multinational companies. Similarly, we are dealing with a communal interest if these decisions are seen to be the full and legitimate responsibility of a limited social community, for example, a specific sociocultural group. Finally, we are dealing with a genuine public interest if the design, realization, and maintenance of the relevant systems has been, or could be, democratically supported by the public.

The way I have phrased these alternatives shows that arguing for the claim that supra-individual interests are public interests includes a normative premise and hence it differs from the argument, provided in the previous section, concerning the existence of supra-individual interests as such.[5] In spite of this, I think that this claim may be, and should be, acceptable for *democratic* individualist liberalists and communitarians as well. After all, as soon as the nonexistence of supra-individual interests has been shown to be a myth, as democrats they cannot consistently delegate the full power to decide on the realization and maintenance of the relevant supra-individual resources either to some private companies or to some specific social communities.

2.2 A further important question is what, exactly, is a *public* interest? Which public is implied in this notion? From a communitarian perspective, the answer is: a particular group, class, or nation. Moreover, communitarian accounts often conceive of such social communities as largely closed and self-contained entities, quite analogous to Thomas Kuhn's notion of paradigms and their associated scientific communities. I think that such accounts are questionable, both in the case of the history of science and in the more general case of the constitution and development of societies (Radder 1996, 183–87). However that may be, the idea of closed and self-contained social collectives, including nations, is certainly inadequate in the case of our present-day societies. For decades, these societies have been influenced by a variety of large-scale interactions with worldwide technological systems, economic policies, political institutions, social organizations, legal arrangements, and cultural interpretations. The realization of socio-material LTSs is just one illustration of this process of globalization.

For this reason, "the public" cannot, and should not, be limited to particular groups, classes, or nations. Instead, it should include all the people who are, or will probably come to be, affected by the issues at stake. In typical cases, this will be a *large* and *indeterminate* number of people, sometimes including the current world population or even some future generations (see Pogge 2001; Gabriëls 2001, 72–78; Offe 2012). Taking into account this kind of public is obviously a big challenge, and hence the notion of a public interest is best interpreted as a *regulative value*. Values point toward general states of affairs that are positively seen to be worth realizing.[6] Even if these states of affairs may never be reached, as is likely in the case of regulative values, such values are significant because they may *guide* debates or policies in a specific direction, which would be different had these debates of policies been guided by quite distinct values.

2.3 Finally, it is important to distinguish *two types of public interest*. The first type concerns an interest in realizing and maintaining positive (or preventing or removing negative) states of affairs of basic significance, which affect all members of the public more or less equally and which are democratically judged to be of public import. Illustrations are the general availability of clean air and safe drinking water, or the global and effective implementation of elementary human rights. Although no endorsement of a public interest will go fully uncontested, interests of this first type can be expected to find fairly wide support. The second type concerns an interest in realizing and maintaining positive (or preventing or removing negative) states of affairs of basic significance, which do not affect all members of the public equally but which are nevertheless democratically judged to be of public import. That is to say, the public may attribute a public-interest status to specific states of affairs that, at a particular time, merely affect a part of it. An illustration is a state of affairs, in which a part of the public suffers real poverty. Yet, even this type of public interest is clearly different from a privatized or a communal interest, since it does *not exclusively* pertain to a particular part of the public. After all, if a quite different group would come to suffer poverty tomorrow, the public interest in alleviating its predicament would apply to them as well. In this sense, the second type of public interest also possesses a certain generality: it applies to *any* particular group to whom it is, *or might become*, relevant.

Again, we see that public-interest claims never denote mere "matters of fact," since they essentially require democratic, normative support. This applies even more to the interpretation of a particular statement of a public interest and to the ways in which it is taken to be advanced. For instance, the questions of what constitutes "real poverty" and how it should be tackled may be matters of serious controversy.[7] For this reason, establishing what will count as a public interest requires the presence of a well-developed and democratic *public sphere*, in which a plurality of views concerning the relevant issues can be addressed openly and critically. Furthermore, it is not enough to endorse the presence of such a public sphere in an abstract way. What is also needed is the institutional implementation of concrete policies that create and maintain the material, social, and cultural conditions required for a free and flourishing public sphere (see Habermas 1998, 244–52; Fuller 2000, 11–17; Fraser 2014). Even so, there is of course no guarantee that a democratic agreement on what counts as genuine public interests will be reached.[8,9]

NOTES

1. I leave the question of the extent to which advocates of these different doctrines are individualist to others.
2. In his wide-ranging study of standards, Lawrence Busch argues that standards are "part of the technical, political, social, economic, and ethical infrastructure that constitutes human societies" (2011, 13), and he concludes that they "display *anonymous* power. Even if we know who established them, standards take on a life of their own that extends beyond the authorities in time and space" (2011, 29). Interestingly, Busch ascribes an important role to neoliberalism in "standardizing the world."
3. For a more extended argumentation, see Radder (1996, chs. 6 and 7); Feenberg (1999, Part II); Brown (2007, 334–36).
4. This is a point that not only individualist liberalists but also some republican political philosophers fail to acknowledge (see the discussion of social inheritance in Fuller 2000, 25–27).
5. Thus, emphasizing that the mediating role of technology differs from advocating the doctrine of technological determinism (for a review of the latter, see Wyatt 2008).
6. Again, we should distinguish between possessed (perhaps better, endorsed) and constituted values.
7. See Gabriëls (2001, 134–83) for an in-depth account of the debate on poverty in the Netherlands during the last decades of the twentieth century.
8. For several caveats concerning what critical deliberation (in a public sphere) may accomplish, see Radder (2000).
9. I would like to thank Mark Brown, Han van Diest, David Ludwig, and Bert Musschenga for helpful comments on an earlier draft of this chapter.

Chapter 3

The End of Trust in the Age of Big Data?

Daniel J. Brunson

Trust is a fundamental aspect of human life, ranging from the trust children have toward their parents to thousands of daily interactions. Indeed, we have to teach children to distrust strangers. Just as our lives are increasingly mediated by technologies that we take on trust, an assessment of how technological mediation has changed our epistemic and existential circumstances is increasingly needed for the sake of raising public awareness. Hardly anyone reads the End User License Agreements and Terms of Service required for the hundreds of apps and websites we use; by one estimate, reading only the privacy policies for websites would take an average person 244 hours a year (McDonald and Cranor 2008). Occasionally, people trust their GPS directions so much that they drive into lakes and swamps, besides to more prosaic accidents. We trust the security of cloud storage to back up our most intimate photos and videos, even with the hacking and distribution of nude photos from hundreds of celebrities in August 2014 ("The Fappening").

Even the distrustful cannot escape. For example, Benjamin Mako Hill has run a personal email server for fifteen years. But, because so many other people use Gmail, his email is not personal: "Despite the fact that I spend hundreds of dollars a year and hours of work to host my own email server, Google has about half of my personal email!" (Hill 2014). Combine this corporate surveillance with the revelations of Edward Snowden about the NSA and its partners, and we have a surveillance state justified by both constant invocations of "terrorism" and customer service. In this way, while corporations and governments tell us to trust them, they no longer have to trust us because they own an increasingly complete record of our daily lives, from our interests, to locations, and even our heartbeats. Moreover, while this sort of health tracking is currently voluntary, corporate and government monitoring of other aspects of our lives is not.

Trust in others allows us to expand and secure our capacities, while also making us vulnerable to betrayal. When we empower others to act on our behalf, we can rarely be sure that their interests will not trump our own. This is called the principal-agent problem, and derives from asymmetrical information, as the one we trust (the agent) knows better than we do whether they intend to fulfill our trust (cf. Sappington 1991). One response to this problem is to remove information asymmetry through an increase in monitoring. However, while monitoring might reduce vulnerability, it is also contrary to trust. Whereas epistemologically a tension may be seen between two views regarding epistemic trust, that is, the goodwill and the risk-assessment views, this chapter explores the tension in light of technological mediation and with respect to such mediation's consequences in terms of social epistemology. In the age of Big Data, we are monitored as never before, and the consequences of the resulting information asymmetry, founded upon technological mediation, may be seen in terms of both the goodwill and the risk-assessment views.

ISSUES OF TRUST

The nature of "trust," like its cognate "truth," is perhaps clear until it is subject to philosophical analysis. We trust our senses, our dear friends, or our furniture, even if they occasionally disappoint us. In this way, trust is often revealed only through its violation—we do not realize whether, or how much, we trust someone until they betray our confidence. And yet, maintaining the possibility of disappointment is a key feature of trust. That is, seeking to minimize the possibility of betrayal through additional *monitoring*, such as reviewing a partner's browser history, is a sign of distrust (cf. Hertlein and Piercy 2006; cf. Hesper and Whitty 2010). In other words, the maxim "Trust, but verify" is perhaps self-contradictory. This is the *first tension* in an account of trust: we trust others to act in agreeable ways to minimize risk, but this trust is itself a source of risk.

Trust seems to involve more than a calculation of risk, for we have warm feelings toward those we trust, and expect reciprocity in such feelings. A friend is someone more than a person evaluated as low-risk, like a coworker we barely talk to but consider "safe" (as expressed in our behavior, perhaps). Instead, it is unlikely that we would characterize our relationship with this coworker as one of trust. This suggests that trust involves more than the absence of distrust, or a favorable risk assessment. That is to say, trust has an affective dimension, involving mutual care and goodwill. Here is a *second tension* in an account of trust: it seems to involve both rational risk calculation and affective stances, but these are opposing approaches to relationships.

For example, I might trust a new car because I have evaluated it to be trustworthy (low-risk). However, perhaps I am making a category mistake, or am speaking metaphorically at best, for a car cannot reciprocate goodwill. On the other hand, there might be some reasonable sense in which my insurance company has goodwill toward me and cares about my well-being. Similarly, Google Now is eager to provide me with the information it thinks that I want, and so in some way cares about my happiness.

The following may be noted in regard to these tensions. First, most accounts of trust take interpersonal trust as the paradigm, and so see other uses such as institutional trust (in a government, say), or even self-trust, as derivative. Furthermore, trust and trustworthiness are correlative, but distinct, in that trust is, at least according to an especially influential view, an attitude, while trustworthiness is a property (McLeod 2011). Within this dominant paradigm, three conditions significant for social epistemology are uncontroversial: vulnerability, thinking well of others, and optimism regarding competence.

1. *Vulnerability*. Trust involves vulnerability to the risk of disappointment or betrayal. This is illustrated in that the efforts to eliminate this vulnerability through monitoring, or through strictly enforced and detailed contracts, actually show a lack of trust. However, trust might involve a particular kind of vulnerability, for many follow Annette Baier in distinguishing between trust and (mere) reliance (Baier 1986, 234–35). That is, while I might legitimately feel disappointment after the failure of something upon which I relied, such as a toaster oven, I should not feel betrayed. Similarly, in cases of monitoring, I would be relying upon someone to a certain extent, but not trusting them. So, when a student submits a paper that a plagiarism-checker finds is indeed plagiarized, feeling betrayed by the said student would be pathological, as the use of the plagiarism-checker shows that I never truly trusted them in the first place.
2. *Thinking Well of Others*. Vulnerability to betrayal rests in part upon a second condition, that is, thinking well of others. For as Govier asserts, "To trust a friend is to believe that her motivations (toward oneself) emerge from affection, care, and concern and not from dislike, ambition, or egoism" (Govier 1993, 104). Further, this condition entails a lack of suspicion, since when we do suspect someone's competency or motives, we engage in monitoring. Hence, consider the following two features regarding trust: first, the exact nature of a trustee's motivation is controversial; the key for this condition is that the trustor thinks that the trustee has an appropriate motivation. Second, located at the intersection of Govier's (1992) work in feminist care ethics and social epistemology, social conditions affect our capacity to think well of others. For example, systematic misogyny or racism, especially when publicly condoned,

reduces or eliminates the possibility of trust across gender and race lines (cf. Sakile and Orbe 2011; cf. Goldberg 2013).

3. *Optimism regarding Competence.* Finally, the third generally accepted condition is that "trust is generally a three-part relation: A trusts B to do X" (Hardin 2002, 9; cf. Zagzebski 2012). In other words, trust requires optimism regarding the competency of the trustee; even if it were possible to "simply" trust someone, which would mean we trust their competency in all (relevant) domains. We should not read this condition too strictly, as in only involving claims like "I trust you to do the laundry." Instead, assessing someone's competency can include affective, cognitive, and moral dimensions. Furthermore, perceived competency would also depend upon the difficulty and importance of the task. For example, I might trust my local mail carrier for most things, but go directly to the post office to purchase tracking and insurance for an expensive or time-sensitive package. Again, here the literature expands to include an analysis of social influences on perceived competency, as in the work of Fricker (2009), Medina (2012), and others on epistemic injustice, critiques of ableism (cf. Ho 2011), and issues of "pragmatic encroachment" in justification conditions (cf. Stanley 2005).

As the above three conditions show, *to trust is to expose yourself to risk.* Moreover, to refuse to trust anyone but one's self and the apodictically certain disclosures of one's own individual rationality may be seen as sawing off the epistemic limb upon which one is sitting. Again, though interpersonal trust may be the dominant paradigm, technological mediation does not eliminate these conditions of epistemic trust.

Whereas the above three conditions regarding interpersonal trust pertain primarily to the first tension noted above, I will focus here on the second tension as what I take to be the prime area of debate in the literature: the nature of trustworthiness. The dominant view derives from Baier, and argues that being trustworthy requires goodwill toward the trustor. That is, not only must the trustor think well of the trustee, the trustee must also care about the trustor, at least in regard to the specific ends of the partnership. This goodwill condition allows us to maintain the distinction between reliability and trust, in that things without wills cannot have good will. For example, my car may disappoint me through its unreliability, but it cannot betray me. Furthermore, this condition captures the sense in which a trusting relationship seems to require a mutual positive affect.

In opposition to the dominant view is a view that prioritizes rational risk assessment: a person is trustworthy when that individual assesses the likelihood of the other person's failing as low. A difficulty for the opposing view is that, though a competent trustee with goodwill may be considered low-risk, the risk-assessment view also permits motives other than goodwill. For example, a competent trustee may be determined low-risk without goodwill if held to the more naturalistic view (e.g., hedonism, egoism). This should be seen as a difficulty for the opposing view, since it may count individuals as trustworthy who act from motivations such as fear of losing their job or competitive aggression.

Though these competing views may be extended across other epistemic debates regarding the nature of trust—is it a belief or an emotion, is it justified internally or externally, etc.—I conclude this section by indicating some issues regarding these views and just the question of epistemic trust. Regarding the goodwill view, I think that its proponents overstate the distinction between trust and reliance. For example, while it may seem pathological to feel betrayed by a microwave, it does not seem pathological to say that I trust my microwave. Likewise, I am comfortable claiming trust in my bank, and can also conceive of circumstances in which I would feel betrayed by my bank, even though I may not know anyone who works for my bank. Though it is beyond the scope of this chapter, two further questions would need to be explored. First, how does the relationship between the parts and wholes pertain to agency and trust (cf. List and Petit 2010/2011), for example, the relationship between the employees of the bank and the agency of the bank itself? Second, is it a limitation of the goodwill view that it might extend "trust" to things that are merely reliable?

The risk-assessment view faces issues as well. First, understanding trust as a rational calculation of risk perhaps neglects the way in which trust is basic. That is, much like the evaluation of testimony, we often lack the capacity and time to perform a full risk assessment in every instance, let alone most of them. This point is at the heart of Coady's (1992) rejection of Humean reductionism in the epistemology of testimony—that we should only trust that testimony which we have personally verified. In addition to perhaps demanding too much from individuals, the risk-assessment view tends to be overly intellectualist, in that it requires neither the trustor's goodwill toward the trustee, nor the reverse. On the one hand, this accounts for the reasonableness of claiming to trust, for example, the reasonableness of a liar to lie. That is, while I would not confide in this person, I trust them to continue to behave as they have. On the other hand, a concern here may be that such a view reduces trust to reliability or bases trust solely upon self-interest. More generally, viewing trust in terms of risk assessment has the difficulty that such approaches to trust-based relations may negatively affect the trust involved, for example, recall the above discussion of monitoring.

Lastly, both the goodwill and risk-assessment views of trust are solutions to the general "principal-agent" problem. In order to achieve our goals, we must rely upon (or entrust) others to act on our behalf. However, there is a fundamental information asymmetry between principals and agents, in that an agent knows more about their capacities and intention to fulfill their charge than the principal does. Thus, we seek to find reliable agents, or use incentives and monitoring to ensure reliability. The classic example is insurance. An insurance company estimates recklessness, charges accordingly, and includes a deductible so that we have "skin in the game." Hence, insurance is the classic example because it seems to adhere to an understanding of trust operable in terms of both the goodwill and risk-assessment views. However, the process of determining insurance rates is, of course, competitive; because I want to receive more in claims than I pay, I might misrepresent my recklessness in order to lower my costs. At least, this is how things worked before Big Data.

BIG DATA AND THE END OF INFORMATION SYMMETRY

Luciano Floridi, philosopher of information and advisor to Google, offers a summary of what it means to live in an age of Big Data:

> A few years ago, researchers at Berkeley's School of Information estimated that humanity had accumulated approximately 12 exabytes [12 billion gigabytes] of data in the course of its entire history until the commodification of computers, but that it had already reached 180 exabytes by 2006. According to a more recent study, the total grew to over 1,600 exabytes between 2006 and 2011, thus passing the zettabyte (1,000 exabytes) barrier. . . . Everyday, enough new data are being generated to fill all US libraries eight times over. (Floridi 2014, 13)

This "zettaflood" of information provides new problems as well as opportunities. For example, the production of data far outstrips our ability to store it in any stable fashion (Floridi 2014, 17–22). This quantitative problem comes with a qualitative one, in that much of this data is of poor quality, or lacks the metadata needed for easy retrieval. Nonetheless, most people emphasize the opportunities of Big Data, such as the unprecedented ability to aggregate public health information demonstrated with Google Flu Trends (Lohr 2015, 107–8; see also Ohm 2013). Of course, there are important privacy issues with the use of personal information, even if "de-identified" and used in the aggregate. Let us set these aside, however, and look at a case where a person volunteers information, as in the case of car insurance.

Car insurance is a form of gambling, in which the insurance company bets that I will pay them far more than I will ever claim. However, my driving record is a poor basis for induction, because having had no accidents might be a measure of luck rather than safety. Hence, statistical analysis is used to help the insurance company determine what kind of person I am to better estimate my recklessness. This is why premiums for male drivers often go down after age twenty-five—statistically, you have "become" a safer driver. Insurance rates have become more nuanced as they've analyzed more data, but as Cowen and Tabarrok (2015) argue, new technologies might end information asymmetry:

> Today, Progressive Insurance offers "Snapshot," a simple device that connects to the [On Board Diagnostic] port. The Snapshot device continuously streams vehicle data such as speed, time, VIN number, and G force to Progressive. Consumers who drive fewer miles, drive during daylight or early evening and without sharp braking typically receive lower rates.

That is, thanks to cheap sensors and wireless communication, Progressive can have constant, real-time data about my driving, and adjust rates accordingly—perhaps even by the mile. While currently a voluntary program, in a decade it could be as obligatory and ubiquitous as CCTV. The key point here is that Progressive eliminates its vulnerability through monitoring—they no longer have to trust us. Similar programs are in place with health insurance through the use of fitness trackers, and now Apple's HealthKit and ResearchKit platforms aim to turn every iOS device into a tool for medical diagnosis and research. On the other hand, we also have more information about companies and products, ranging from comparative pricing to crowd-sourced reviews such as Angie's List and Yelp. This serves to minimize our risk of buying a bad microwave or seeing an unpleasant doctor.

Nonetheless, companies and governments have access to more data than we do, as well as the ability to process it. Another use of Big Data, and the accompanying reduction of information asymmetry, is the reduction of moral hazard, "the tendency of a better informed party to exploit its information advantage in an undesirable or dishonest way" (Cowen and Tabarrok 2015). For example, having car insurance can lead to more reckless driving, as the insurance will cover the costs of an accident. Instead, real-time monitoring allows for greater manipulation of our "choice architecture, meaning the background against which choices are made. Such architecture is both persuasive and inevitable, and it greatly influences outcomes, whether or not we are even aware of it" (Sunstein 2014, 14). Sumptuary, or "sin" taxes, are a common example of this sort of manipulation, as we seek to discourage undesirable behaviors like smoking through increasing the direct cost. Libertarian paternalism (Thaler and Sunstein 2008/2009; Sunstein 2014) argues that it is both moral and necessary to "nudge" people to make choices in their best interest, given the empirical evidence of our "bounded rationality" (Kahneman 2013). This mindset, combined with real-time monitoring and feedback, allows for the gamification of our daily lives—motivating preferred behaviors through achievements and points. For example, a fitness tracker might determine that I often fail to go to the gym after work and so award bonus points to inspire me to go more often. This is a carrot, while knowing that my doctor and insurance company have access to the record of my heart rate is a stick. In a way, then, not only does my doctor no longer need to trust me, but I no longer need to trust myself, as an algorithm will decide how best to optimize my behavior.

THE AGE OF ALGORITHMIC AGENTS

Danaher (2014) coins the term *algocracy*, rule by algorithm, in his analysis of Morozov's (2013b) argument about the real threat to privacy in the age of Big Data. First, as we saw above, many people rely upon algorithms to make decisions, such as navigating to an unknown destination. Second, in response to the "zettaflood" discussed by Floridi, many people happily subject themselves to content customization, relying upon Amazon, Google, etc., to provide advertisements, entertainment, and news that appeals to their previous patterns. The concern here is that the internet has become a small set of echo chambers, in which background processes filter out challenging or random information in the name of customer service. Third, and more in line with Morozov and Danaher's arguments, is that the advances in machine learning made possible by Big Data makes these algorithms opaque even to their designers. Machine learning involves a variety of methods, but the basic point is that programs develop new programming based upon experience (Mitchell 1997). This self-programming might concern new means to an externally given goal, or the determination of new goals. While still in its infancy, the concern is that these algorithms have become too complex for humans to understand. Furthermore, algorithmic agents make decisions that affect human life millions of times a day, from flagging "suspicious" credit card purchases, to identifying "terrorist" internet activity, and making the majority of stock market transactions. On May 6, 2010, the US market experienced a "Flash Crash" where in a period of about thirty minutes the Dow Jones Industrial Average dropped nearly 1,000 points and then rebounded. While the cause of this crash is still unclear, "what's for certain, however, is that the market couldn't have moved so far so fast if algorithms, which act independently of humans and require less than a second to place and complete a trade, didn't own the market. But they do" (Steiner 2012, 5).

These abstract algorithmic agents might be poor candidates for trust, understood as more than reliance. However, programs such as Siri and Cortana strive to be personal assistants, and as they develop more personality might qualify for trust. Of course, they have limited competency, but that too will improve with more data. Perhaps they, or their descendants, will never possess the affective component considered essential by the goodwill view of trust, but their behavior will likely be functionally indistinguishable from a caring friend.

CONCLUSION: THE END OF TRUST?

One end, or purpose, of trust is enabling us to achieve goals beyond our individual capacities. Whether we achieve this end through mutual goodwill or rational risk mitigation is an open issue, and perhaps we do not need to reconcile or reduce these accounts. What is clear, though, is that both face new questions in light of the age of Big Data. For the goodwill account, the increasing use of algorithmic agents, both impersonal and eventually personal, suggests either goodwill could become a minimal element of our lives, or we will need to reconsider what sorts of agents might possess goodwill. For the risk-calculation view, the promise of ubiquitous real-time monitoring suggests that risk assessment will become increasingly accurate. However, these risk assessments are already incomprehensible to individuals, and so rather than reducing the need for trust, the burden shifts to the corporations and governments that monitor us; or, rather, the algorithms that they employ. But as the "Flash Crash" or multiple instances of hacking personal data have shown us, perhaps algorithms make us susceptible to catastrophic risks. In these ways, and others, our changing social conditions call for reconsideration of our fundamental epistemic and moral concepts.

Chapter 4

Filter Bubbles and the Public Use of Reason

Applying Epistemology to the Newsfeed

Jamie Carlin Watson

Part of answering the social epistemological question *How should the pursuit of knowledge be organized?* involves identifying obstacles to the pursuit of knowledge and developing strategies for overcoming them. There is some reason to believe that the technology intended to enhance our access to evidence has the unintended consequence of constraining that access. Even worse, this technology makes it easier for a small minority to manipulate our beliefs and interests in invisible ways. Here, I explain a phenomenon called *filter bubbles* and how the preferences we express through our use of the internet can be used to undermine epistemically responsible behavior. I argue that what I will call *filtering bias* is a type of censorship in which there is significant public interest and that, although there are no promising solutions to the problem, understanding this obstacle can help us guard against complacency when using technology as part of our epistemic activities. Thanks to Frank Scalambrino and Laura Guidry-Grimes for helpful comments on earlier drafts.

FILTER BUBBLES

A classic liberal concern about censorship is that ideas necessary for knowledge or well-being can, through prejudice or error, be suppressed by authorities with a monopoly on the use of force. Importantly, anticensorship laws do not protect citizens from *private* restrictions on speech—for instance, in political or religious organizations—because doing so would require the state legally to prevent such groups from preferring one set of ideas to another, thus making the state's behavior indistinguishable from direct censorship. The idea that all perspectives should be expressible in every context is itself an idea that might threaten knowledge or well-being. Anticensorship laws do, however, require the state to protect citizens against other citizens who would restrict, whether by force or intimidation, the freedom to speak publicly. Rather than taking a stand on any particular idea, the purpose of the state is to preserve and protect fairness in the expression of ideas, not fairness in the ideas themselves.

In general, biased speech and speech that attempts to persuade are not regarded as attempts to restrict speech because citizens are not prevented from rejecting such content or seeking alternative, contrary perspectives. Further, those who disagree are free to challenge publicly the veracity of this content and its sources. And these conditions may remain in place regardless of whether the attempts to persuade or respond are fallacious or cogent, whether they are disingenuous or well intended. Thomas Jefferson, for example, argued that the state should not restrict public expressions of speech because whoever decides which speech to restrict "will make his opinion the rule of judgment, and approve or condemn the sentiments of others only as they shall square with or suffer from his own" and because "truth is great and will prevail if left to herself."[1] The playing field for ideas may remain fair enough even in heated, rhetorically filled banter so long as others are not prevented—by threat or intimidation or force—from publicly objecting.

Unfortunately, dramatic shifts in the history of ideas and the discovery of pervasive cognitive biases suggest that Jefferson was overly optimistic about truth's chances. Truth may never be among the perspectives expressed or debated.[2] Happily, liberalism allows that, even under unfavorable conditions, we may still *pursue* truth by trading in the marketplace of ideas, subjecting new, innovative, and controversial ideas to public scrutiny without fear of legal sanction. We may still act in epistemically responsible ways. The result might not be truth, but there is ample evidence that such freedom facilitates progress and allows us to experiment with who we are as individuals and cultures.

Perhaps ironically, some of the technological by-products of the free exchange of ideas are starting to threaten our ability to pursue truth. Though the internet promises almost unlimited access to ideas, humans cannot sort and comprehend vast amounts of information. Algorithmic filtering tools are necessary if these sources are to be a useful and efficient means of acquiring information. Algorithmic filtering is one way of easing the effects of information overload. It tailors the vast amounts of information on a topic to that which best fits what the data collectors know of us.

To be sure, filtering of one sort or another has always occurred. There is too much news to report, so news outlets must decide which stories are hot, interesting, or worth reporting and reject the rest. They must also frame stories in a historico-political context, which presupposes a particular understanding of history. They must choose how much detail to explain. For example, a recent news item is that US president Obama proposes reducing tariffs to increase

American jobs. Since basic economics tells us that lower tariffs typically mean outsourcing domestic labor, it isn't clear how lower tariffs will *increase* domestic jobs. Critics have highlighted this worry, yet neither the *New York Times* nor *Huffington Post* explains the economic relationship between tariffs and domestic jobs.[3]

Internet filtering tools offer even greater scope and precision in the filtering process, which raises questions about how the algorithms should be designed. Since liberal democracies reject the idea that a state-sponsored mechanism can effectively or justly filter, we are obliged to look for alternatives. The results would seem to be no better if private individuals (such as those at particular big media outlets)[4] make the filtering decisions, so the most plausible option seems to be the one based on our individual interests—markets and the information we volunteer about who we are. And this is precisely what has happened.

Algorithms developed by Amazon, Google, and Facebook paved the way for data gathering through free services like Gmail—where users volunteer personal information in exchange for an email account—and *click signals*—the data we produce when we use the internet: the articles we search for, the length of time we view them, those we share, those we *like* on Facebook, the types of devices we use to view them, the locations from which we view them on our devices, etc. In his book *The Filter Bubble*, cultural critic Eli Pariser explains that "by getting people to log in, Google got its hands on enormous piles of data—the hundreds of millions of e-mails Gmail users send and receive each day. And it could cross-reference each user's email and behavior on the site with the links he or she clicked in the Google search engine."[5]

Click signals are recorded and mined for patterns that allow producers to cater to users' particular interests. This is why the shoes you were shopping for on Zappos now appear in an ad on your Facebook page. It also explains why one person's Google search looks very different from another's,[6] and why two people with many of the same Facebook "friends" see very different posts in their newsfeeds—they *like* different posts, and subsequent posts are filtered according to those preferences.

The data gathered on internet users is now staggering, and companies like Acxiom store and sell this information to companies. Acxiom has information about "96 percent of American households and half a billion people worldwide," including "the names of their family members, their current and past addresses, how often they pay their credit card bills[,] whether they own a dog or cat (and what breed it is), whether they are right-handed or left-handed, what kinds of medication they use (based on pharmacy records) . . . the list of data points is about 1,500 items long."[7] This information is for sale to anyone willing to pay, and it is available to the federal government on demand.[8] All this raises significant moral concerns about privacy. But what about epistemic concerns? How does this data gathering affect our ability to pursue truth?

A result of filtering for individual interests is what Pariser calls the "filter bubble." Filter bubbles occur when the information you volunteered to websites, along with your click signals, leads to an isolation bubble of information that almost perfectly suits you. "We're never bored. We're never annoyed. Our media is a perfect reflection of our interests and desires."[9] If you regularly visit conservative websites and *like* conservative posts, the information you see from nonconservatives (say, liberals and libertarians) will diminish or disappear. Given that the world looks conservative (or liberal, or libertarian) to you, it will be difficult to take an opposing point of view seriously.

So far, though, these are not new concerns about epistemic responsibility. People have always been free to stick their heads in the sand. The problem arises when we combine the invisibility of the filter bubbles with the evidence that humans have behavioral tendencies that can be manipulated by others. We are faced with the question of whether filtering algorithms shape content so much that companies and governments can engage in what I will call *indirect censorship*, an invisible structuring of our world such that we see primarily what someone else wants us to.

Using your personal information to invisibly influence future preferences and purchasing behavior is called *persuasion profiling*.[10] Pariser explains, "In the wrong hands, persuasion profiling gives companies the ability to circumvent your rational decision making, tap into your psychology, and draw out your compulsions. Understand someone's identity, and you're better equipped to influence what he or she does."[11] Based on what we know about human psychology, corporations and internet barons can use our personal information to manipulate and change our interests, even if not from malicious motives. This sounds very much like citizens restricting other citizens' ability to pursue truth by selectively structuring the way they receive evidence. If this is right, there is a public interest in the phenomenon of filter bubbles that has both moral and epistemic dimensions. The question I want to address is: *How can we be epistemically responsible if our online world is manufactured by others?*

MEDIA BIAS VERSUS FILTERING BIAS

To begin, we must tease apart the moral and epistemic concerns associated with this sort of indirect censorship, though they overlap at a crucial point, as I argue later. Morally, censorship threatens autonomy and the values associated with personal development,[12] and much has been written on this topic. My concern is epistemic. I will offer reasons for thinking that filter bubbles can be constructed in ways that constitute a constraint on the free exchange of ideas that amounts to a kind of censorship and that this constraint challenges our ability to function well as epistemic agents, that is, it encourages intellectual passivity and a false sense of intellectual responsibility.

To clarify the epistemic problem, we must dissociate it from another epistemic concern about media. One might think that the problem with filter bubbles is about the *reliability* of the media that consumers receive through filtering algorithms. For example, some individuals' interests may lead filtering tools to select traditional "big media" sources, such as CNN or NPR, while others may lead to a selection of fringe websites. If one source is less reliable than another and consumers cannot adequately distinguish their reliability, filtering may inadvertently impede our ability to form beliefs responsibly.

Teachers are familiar with the difficulty that students have in distinguishing academic and nonacademic, primary and secondary, sources when writing papers. Further, there is evidence that confirmation bias and herding behavior permeate the voting public.[13] But some argue that, even if media consumers cannot identify reliable sources on their own, as long as *someone* is making sure that sources are reliable, citizens will have access to the information that is necessary for epistemically responsible behavior. And so, a popular debate is whether internet sources that purport to be informational, particularly blogs, are as reliable as traditional media.

Richard Posner (2005) and Alvin Goldman (2008) argue that traditional media is in principle more reliable than personal blogs. Goldman identifies several advantages of the former over the latter. Traditional media involves multiple gatekeepers that serve to vet the merits of an idea—trained news reporters, editors, editorial boards, popular demand, and the individual consumer. In contrast, blogging places the full burden of gatekeeping on the individual. And individual consumers are poor at distinguishing reliable from unreliable sources.[14] Unlike blogging, traditional media often requires more than one source and that the story be checked by fact-checkers before publishing it.[15] Combined, these two reasons to prefer traditional media over blogs suggests a third: the filtering processes of traditional media aim at determining truth rather than promoting a particular perspective. If a prominent news outlet's story is drastically different from all other reputable outlets, its credibility will be called into question. They have incentives to express the very best information.

In response, David Coady (2012) argues that blogs do not actually suffer from these deficiencies. He argues that there is little reason for thinking that bloggers are less concerned with truth than professional journalists, citing the history of pamphleteers and citizen journalists. Professional journalists are often as concerned with appeasing audiences, corporations, and politicians as they are with truth, and some journalists maintain flourishing careers even after being reckless with sources and reporting falsehoods.[16] Further, Nicholas John Munn (2012) argues that, even if blogs did suffer from the deficiencies that Posner and Goldman cite, since 2008 the blogosphere has been so absorbed into mainstream media that most are no less reliable than traditional media.[17]

But even if Coady and Munn are correct that web-based information is not generally less reliable than traditional media, it remains the case that both traditional media and blogs are biased in certain ways, and if filter bubbles constrain our searches according to our interests, we will be exposed only to those sources that confirm our perspectives. Consider how often you look beyond the first three to five hits on a Google search. If you are unaware of the mechanics of filtering, you may think that these are the five most relevant hits for your search. What else are search engines for? In fact, your results are the most relevant *given* the popularity of the site, *given* which websites paid to be at the top of search results, *given* your previous searches, *given* where you are in the world, *given* the device you are using, and *given* personal information gleaned from companies like Acxiom. Google can use up to fifty-seven different pieces of information that can be obtained from every click on the Web.[18]

The epistemic concern is not that internet content is less reliable than other types of media. It is that, while the biases of traditional media are explicit—people are generally aware that Fox News is biased in certain ways and that NPR is biased in others—filter bubbles occur individually, practically invisibly, and involuntarily.[19] Pariser explains that, while a cable channel caters to your interests, it also caters to the interests of all its viewers, so there are gaps in how well it can tailor its information to you. There are no gaps online. It is practically invisible because, while you may know MSNBC's bias, you don't know Google's or Facebook's, and you don't know what information about you they think is relevant for deciding what to filter. It is involuntary because, while media biases change when you switch the radio from Rush Limbaugh to *All Things Considered*, they don't change when you click to a different website; your click signals follow you around the Web, tailoring information to an algorithm's picture of who you are. Alternative perspectives become less and less frequent in your search results.

Imagine that all billboards along the highway can electronically read your click signals. As you drive along you notice the billboards advertise only products and services you have used in the past two months. During election season, those billboards display ads only from your political persuasion. We do not allow corporations to monopolize markets because this undermines competition. We would not allow one political candidate to take down all his opponent's ads and signs. This would threaten the fairness of an election. And yet online, those who control the algorithms have the power to do something very similar. I will use the phrase *filtering bias* to refer to algorithmic media filtering that constrains consumers' evidence-gathering behavior. Apart from moral concerns with fairness, filtering bias threatens our ability to act as epistemically responsible agents.

EPISTEMIC RESPONSIBILITY

The starting point for many theories of knowledge is twofold, suggesting that a person knows a proposition, *p*, if and only if she has a *justified true belief* that *p*, and justification is understood as a matter of a subject's having *good evidence* that *p* is true. Yet it turns out that it is pretty easy to have a justified true belief just by luck, despite that most philosophers reject the idea that you *know* something just because you were lucky. Further, philosophers largely agree that if a belief-forming process or source of evidence is biased toward something other than truth, it is not a source of knowledge. Even if a source were inadvertently to produce all true beliefs, it is only accidental that it does so, and therefore, simply a matter of luck. If its relationship to truth is *seen* to rest on luck, one may resist the temptation to form beliefs on the basis of that source. If the bias goes unrecognized, epistemic agents will be at a disadvantage if the source ceases to produce true propositions.

We understand the epistemic effects of information bias when filters are visible. For example, members of extremely liberal or extremely conservative institutions often recognize that they are shielded from the expression of other perspectives, though they may think that they understand these perspectives because they are explained by respected members of their communities. It takes effort to burst such bubbles and actually read alternative views for ourselves. The longer we put off making the effort, the easier it is to accept what we hear as reliable.

But this is a typical problem for forming responsible beliefs. Filter bubbles present a unique challenge because they reduce the sense that there is a bubble to burst. We put off the effort because filtering algorithms make us feel like we have fulfilled our epistemic responsibilities; they keep us from knowing what we don't know.

Epistemic responsibility involves more than seeking truth and relying on evidence; it also involves the attitudes and dispositions with which we form beliefs. A teenager raised in an extremely liberal or conservative household may genuinely seek truth and may hold justified political beliefs on the basis of all the evidence to which she has access. But if all that evidence is biased toward her family's political orientation, she may not have an accurate sense of the political landscape. She will not realize that she is lacking evidence until she is disposed to look for *all* available evidence. To achieve this more robust sense of the political, she must develop certain epistemic dispositions, such as intellectual humility and courage, open-mindedness, and carefulness. These dispositions are difficult enough when biasing factors are visible. Invisible biasing factors threaten even our willingness to pursue these dispositions to any great extent; they convince us that we have done all we need to do to believe responsibly. We become irresponsible without realizing it.

FILTERING BIAS AND CENSORSHIP

So far, we have seen that filter bias poses a unique threat to our ability to form beliefs responsibly. Since a small group of citizens is responsible for this bias, intentionally or unintentionally, we have reason to worry that filtering bias is tantamount to a politician's systematically removing all her opponent's campaign ads. It suggests that there is unfairness in the epistemic process. And this unfairness looks like a certain type of censorship.

Philip Cook and Conrad Heilmann defend a model of censorship that distinguishes among public and private censorship and public and private self-censorship.[20] Classically, censorship is a restriction on public speech by external censoring groups (either state or social). Cook and Heilmann argue that self-censorship refers to citizens' personal restrictions on their speech behavior. If a state restricts speech and citizens obey, this is public self-censorship. If an extremist group threatens violence if some ideas are expressed (e.g., the violence threatened and perpetrated in protest of images of Mohammad in the Danish newspaper *Jyllands-Posten* and the French magazine *Charlie Hebdo*) and citizens obey, this is also public self-censorship. If a person restricts her own expression because her ideas are inconsistent with the general ethos of a culture or because she is afraid of a particular public reaction, this is private self-censorship.

Complementing Cook and Heilmann's work, we can draw an additional distinction between *direct* and *indirect* censorship. Direct censorship involves the explicit imposition of laws or threats for speaking in a certain way. Indirect censorship involves those behind-the-scene behaviors that prevent certain types of speech. This is usually a function of culture that emerges as a type of implicit bias. For example, decades after World War II, the central descriptive term for the Nazi extermination of the Jews was "Holocaust," from the Greek *holokauston*, which means "burnt offering" in a religious ceremony. The idea that the extermination was an act of sacrifice is offensive, but its popularity among scholars led, albeit unintentionally, to restricting the preferred term, "Shoah," which means "catastrophe." There are many such controversies, for example, about who gets to name a group or a culture. Terms like "negro," "Oriental," and "Mohammedan" were chosen by people who were not members of the groups that they intended to identify. More recently, there was controversy over what to call the terrorist organization ISIS/ISIL/IS(?), given that at least one of these presupposes political legitimacy.

Technology now allows us to embed this type of cultural conditioning into our evidence about the world. The memes and articles we read shape, at an individual level, how we view and talk about those who disagree with us. And people in control of the algorithms can influence that process to suit their interests. Because it is coercive, it is plausible to regard filtering bias as a form of *indirect public censorship*. To the extent that advertisers and politicians are invisibly nudging our worldviews in their favor, they are restricting our abilities to see opposing views through the eyes of those who espouse them. It is here that the moral and epistemic concerns overlap. In protecting citizens from the moral harms that we might inflict on one another by preserving fairness in our public expression, the state also preserves something of our ability to believe responsibly.

PRESERVING RESPONSIBILITY

If all this is right, how should we respond to filtering bias? Given its close affinity with indirect censorship, state intervention seems a natural solution. But in this case, state intervention is not likely to be successful. In addition to its poor track record at regulating markets (from corn subsidies to OSHA), technology is constantly changing, and regulations are almost impossible to revise. Legal loopholes and technological advancements often outstrip sluggish legislative processes. One type of state solution that Pariser thinks might work is to require companies to give users "real control over our personal information." He says the concern should not be on anticensorship laws, but on property rights in our identities (cf. Fuller and Lipinska 2014). [21] But moral identity is a tricky philosophical issue, and such rights are difficult to define. Further, when we are online we often sign over our rights to our images and ideas in agreeing to a site's terms of use.

Individually, we can change our search habits, adding and subtracting websites from our usual patterns from time to time, erasing our cookies, and using sites that give us more control over how our data is used.[22] The problem is that, to take advantage of these strategies, individuals must be aware of filter bubbles. But this lack of awareness is a large part of the problem. Only a certain demographic of the population will be concerned with the blog websites that explain filtering bubbles.

What each proposed solution reveals is that epistemic responsibility is an essentially individual matter. In 1784, Immanuel Kant warned that laziness and cowardice are the primary reasons that people choose not to think for themselves. He argued that political action cannot alleviate this problem. Even in cases of revolution, "new prejudices will serve, in place of the old, as guidelines for the unthinking multitude."[23] The difference now is that we face more than laziness and cowardice; even if we overcome those, we face the manipulation of the very behaviors aimed at enlightenment.

The key point to take away is that, though the internet offers almost limitless access to information, it comes with all the problems that attend masses of data that must be contextualized and interpreted. No one can do the difficult work of believing well for us. Even filters that reflect our complex personalities cannot lay open the world as it is in itself. We must guard against complacency; technology can facilitate enlightenment or it lead our heads deeper into the sand.

NOTES

1. Jefferson, "Bill Establishing Religious Freedom," 1779, *Founders Online*: http://founders.archives.gov/documents/Jefferson/01-02-02-0132-0004-0082.

2. Consider the sixteenth- and seventeenth-century debates over gravity and atoms.

3. Dave Johnson, "Obama to Visit Nike to Promote TPP. Wait, *Nike*? Really?" *Huffington Post*, May 7, 2015: http://www.huffingtonpost.com/dave-johnson/obama-to-visit-nike-to-pr_b_7233118.html; Peter Baker and Julie Hirschfeld Davis, "Obama Chooses Nike Headquarters to Make His Pitch on Trade," *New York Times*, May 7, 2015: http://www.nytimes.com/2015/05/08/business/obama-chooses-nike-headquarters-to-make-his-pitch-on-trade.html.

4. The idea here is that a company like Time Warner or Viacom would make *their* opinions the rule of judgment, which is, arguably, no better than the state. For an example of how corporations might filter poorly, see the documentary *This Film is Not Yet Rated* (IFC 2006), http://www.imdb.com/title/tt0493459/.

5. Eli Pariser, *The Filter Bubble: What the Internet is Hiding From You* (New York: Penguin Press, 2011), 33.

6. Following Pariser's example, I conducted an informal search, asking Facebook friends of various political and religious persuasions from around the world to search "net neutrality" and "media bias" on Google. The results are telling, though not radical. Wikipedia came up first on almost every search, suggesting that popularity still drives the filters. The first two to three hits were the same on almost everyone's searches, from Bolivia to Korea to the United States. The results of those who used more ad blockers and security measures were much the same after the first two or three hits. But for those without such measures, the rest of the hits varied significantly.

7. Pariser, *The Filter Bubble*, 43.

8. "According to one *Wall Street Journal* study, the top fifty Internet sites, from CNN to Yahoo to MSN, install an average of 64 data-laden cookies and personal tracking beacons each" (Pariser, *The Filter Bubble*, 6). The study Pariser references is

Julia Angwin, "The Web's New Gold Mine: Your Secrets," *Wall Street Journal*, July 30, 2010: http://www.wsj.com/articles/SB10001424052748703940904575395073512989404.

9. Pariser, *The Filter Bubble*, 12.

10. Pariser attributes the phrase "persuasion profile" to Dean Eckles, from a phone interview (121).

11. Pariser, *The Filter Bubble*, 123. The phrase "in the wrong hands" is important, since some argue that persuasion profiling can benefit consumers. As an example, Pariser cites DirectLife, "a wearable coaching device by Philips that figures out which arguments get people eating more healthily and exercising more regularly" (Pariser, 121). The popular *nudge* literature also advocates using persuasion profiling in ways that benefit individuals according to their own sense of what is beneficial (cf. Thaler and Sunstein, 2008/2009).

12. Prussian philosopher Immanuel Kant suggests that legally enforcing one set of ideas impairs our ability to gain new insights and root out errors, and he takes this curiosity as part of what it means to be human; it "would be a crime against human nature whose proper destiny lies precisely in such progress" ("What is Enlightenment?" 1784, trans. Mary C. Smith).

13. Thomas Gilovich, *How We Know What Isn't So* (New York: Free Press, 1993); Thaler and Sunstein, 2009.

14. Goldman, Alvin. 2008. "The Social Epistemology of Blogging." In *Information Technology and Moral Philosophy*, edited by Jeroen van den Hoven and John Weckert. Cambridge, UK: Cambridge University Press, 8.

15. Ibid., 7.

16. David Coady, *What to Believe Now: Applying Epistemology to Contemporary Issues* (Malden, MA: Wiley-Blackwell, 2012), 151–57.

17. Nicholas John Munn, "The New Political Blogosphere," *Social Epistemology* 26, no. 1 (2012): 55–70.

18. Pariser, *The Filter Bubble*, 2.

19. Ibid., 9–10.

20. Philip Cook and Conrad Heilman, "Censorship and Two Types of Self-Censorship" *LSE Choice Group*, The Centre for Philosophy of Natural and Social Science 6, no. 2 (2010): 1–5; Cook and Heilman, "Two Types of Self-Censorship: Public and Private," *Political Studies* 61 (2013): 178–96.

21. Steve Fuller and Veronika Lipinska, *The Proactionary Imperative: A Foundation for Transhumanism* (New York: Palgrave Macmillan, 2014).

22. Pariser, *The Filter Bubble*, 222–29.

23. Kant, "What is Enlightenment," 1784.

Chapter 5

The Internet and Existentialism

Kierkegaardian and Hegelian Insights

Patrick J. Reider

In this chapter, I develop existentialist themes that highlight the manner in which the internet can pose a threat to personhood.[1] I begin by expanding upon Herbert Dreyfus' appeal to Søren Kierkegaard (the father of existentialism) as a means to critique the internet. While I support and defend Dreyfus' key claims concerning Kierkegaard and the internet, I argue that Kierkegaard's failure to appreciate the role of social mediation significantly limits his account of agency, and that this limitation becomes especially relevant in considering the relationship between agency and the internet. I further argue that Hegel's conception of personhood can offer corrective measures to the internet age, in which the relation between knowing, as a form of social mediation, and effective agency is obscured. In this vein, I argue that even from an existentialist perspective, which rejects the existence of innate value, social epistemology, as the view that knowledge is irreducibly a social phenomenon, ought to underpin one's search for meaningful action.[2]

1. IDLE ENTERTAINMENT, REFLECTION, AND CURIOSITY

When a question is posed and the answer is unknown, a smartphone typically glides into someone's palm. This often occurs before the intention to "search" for an answer is registered in one's awareness. When this reflexive response fails to produce the answers sought, online groups that harness the collective cognitive abilities of many users are a mere click away.

Futurists would have us believe that these resources allow us to transcend the prior (though often unnamed) gold standard of learning. "Let us now embrace the platinum standard of the internet age—a world of hyper-connectivity and readymade access for all!" seems to be the subtext of new media enthusiasts. Internet proponents assume that this techno-social phenomenon spurs the desire to know, while simultaneously providing the means to know. Moreover, this marvel of modern technology indiscriminately serves the disadvantaged, poor, and marginalized (assuming that they have access to it), as readily as the rich and powerful. Cast in this neoliberal light, is not the existence of such widespread epistemic developments an optimistic sign for our communities, country, and perhaps the world?

In *Anonymity versus Commitment: The Dangers of Education on the Internet*, Dreyfus "translates" Kierkegaard's criticisms of the "press" and its readers to that of the internet and its users. Written in 1846, Kierkegaard's "The Present Age" characterizes his culture as one of idle "reflection and curiosity." Kierkegaard's essay argues that the reflection and curiosity of his age were marred by an unfocused and self-indulgent nature. He proposes that these pejorative aspects of reflection and curiosity largely stem from the popularity and wide distribution of the press, which he believed had the effect of neutering distinctions of value.

In considering Dreyfus' comparison of the "press" in Kierkegaard's time to the internet in our own, we are invited to consider anew the questions that Kierkegaard raised and to see if the change of media has done anything to reduce the relevance of his argument. We can begin by asking whether internet users might fairly be characterized as "driven by an unfocused and self-indulgent curiosity and not a curiosity which reflects an *eros* (i.e., desire) for knowledge." And here we need to consider how the internet is used. Although there are many uses of the internet, it is plausible to argue that its most pervasive use is for entertainment. For Dreyfus, this puts internet usage squarely within the scope of what Kierkegaard was criticizing:

> Commitment to information as a boundless source of enjoyment puts one in what Kierkegaard calls *the aesthetic sphere of existence*—his anticipation of postmodernity. For such a person just visiting as many sites as possible and keeping up on the cool ones is an end in itself. The only qualitative distinction is between those sites that are interesting and those that are boring. (Dreyfus 2013, 643)

Given this assumption, the average internet user's curiosity can be characterized as "self-indulgent" to the degree that it is put to the service of idle entertainment.

The pursuit of electronic entertainment is now inculcated early on as a way of life. Consider, for example, the unforeseen outcome of electronic tablets. These devices are now surrogate babysitters, as parents encourage the distractions of mobile connectivity and an endless variety of videos and games. The instantaneous gratification of jumping from one short video to the next can captivate even the limited attention span of toddlers for hours. This approach continues in formal education, where teachers administer games in which students are challenged to see who can first obtain an answer to a question via their smartphones. The implication is that education merely requires acquisition of information and that such acquisition is fun.

Idle curiosity that seeks entertainment cares not for the truth, which is typically difficult to establish. In this vein, Dreyfus aptly quotes Kierkegaard: "It is frightful that someone who is no one . . . can set any error into circulation with no thought of responsibility" (Kierkegaard 1970, 481). This occurs because the internet, like the press, creates anonymity. One does not have to face, know, or interact with those that one misinforms or slanders. The lack of proximity and interaction with those that one may, intentionally or not, mar or mislead frees one from accountability. Hence, the internet has to a large extent become a "free for all" regarding what ought to be liable acts.[3]

One can typically post whatever one wishes without repercussions. Even disseminating hatefully dishonest content or uploading one's disrobing in the privacy of one's own home tends to have no negative repercussions for those who post it, and less so for those who propagate it. Such acts are committed from the comfort of a parent's sofa, the sleepy spouse's tablet before retiring to bed, and in heated leather seats of luxury cars. In this manner, internet users are lulled into a false sense of inconsequentiality concerning their actions, because they are typically performed under conditions of great ease and comfort. However, once such convenient acts of slander, invasion of privacy, or sexual abuse are uploaded, they tend to become permanent features of the Web. These actions are doubly disconcerting when one considers that adolescents and young adults are among the most common perpetrators and victims of such crimes.

2. KNOWLEDGE AND THE INTERNET

The ease by which internet users obtain information, even highly specialized information, has a tendency to undermine one's openness to the fact that mastery and nonsuperficial knowledge require hardship and failure.[4] It is popularly felt among my students that, in this day and age, one is entitled to forego hardship, even that of rigorous academic challenges. This popular belief, as it concerns academic success, seems to stem from the manner in which people often place high value on the accessibility of information but little value on the hard-won skill of discerning the true from the false, relevant from irrelevant, and skilled assessment from lazy opinions. This epistemic softness is a cultural pandemic.

Seeking information for the sake of entertainment fosters the habit of being unfocused. In order to acquire the rich benefits of knowledge, however, one needs both focus and purpose. Purpose permits focus by guiding the process of forming useful questions, permits the discrimination of relevant content from that which is not, and thereby promotes applicable and beneficial outcomes.

Technologically entrenched societies seem to be unaware that focus is a learned trait, whose acquisition requires an unpleasant strain. It should thus be obvious why those cultures that foster an "on-demand" diet of cable television, streaming video, handheld gaming, texting, and social media find focused activity difficult and undesirable. It should be equally obvious that epistemic incompetency must surely follow in that one fails to develop the ability to judge successful instances of support from unsuccessful instances.

As already noted, the use of the internet for the satisfaction of curiosity quickly devolves into a form of entertainment. Since the pursuit of information for the sake of entertainment lacks purpose beyond that of immediate satisfaction, it fosters a disinterest in truth. Without focused discipline and purpose, one is apt to believe that the hoarding of colorful baubles of disparate details equates to the acquisition of knowledge. The common perception that one can readily obtain accurate answers to just about anything they wish to know by casually browsing the internet fosters this belief. And this in turn foster epistemic incompetency.

When a person lacks defining commitments or internal continuity of purpose, they lose the capacity to make distinctions of lasting value that transcend momentary desires. This leads to "leveling," in which every kind of behavior, every sort of achievement, and every type of goal become relativized until nothing seems good or bad beyond how it makes one feel. According to Kierkegaard, leveling throws one into "despair." In this context, "despair" concerns a state in which all values seem to dissipate. When one is captured by despair, personal choice seems futile, for no worthy purpose appears adequate to guide one's choices. This has serious consequences for knowing as a committed pursuit.

3. THE EXISTENTIAL THREAT OF THE "PUBLIC"

According to Dreyfus and Kierkegaard, a dangerous and unintended outcome of having a press that widely distributes information on nearly every topic available is the rise of do-nothing commentators: "The Public sphere . . . promotes ubiquitous commentators who deliberately detach themselves from the local practices out of which specific issues grow and in terms of which these issues must be resolved through some sort of committed action" (Dreyfus 2001/2009, 75). Hence, the Age of Enlightenment, in which individuals are believed to be empowered because of their freedom to rationally discuss social issues, unintentionally promotes a culture that avoids commitments. This precedence is set by the press: "What seems a virtue to detached Enlightened reason" is actually discourse in which "even the most conscientious commentator doesn't have to have first-hand experience nor take a concrete stand" (Dreyfus 2001/2009, 76).

The press produces discussion that makes it appear as if individuals are actually taking a stance such that they can be unified by shared practices and commitments. But these perceptions are produced by abstraction, and create the illusion of what Kierkegaard calls the "public." For Kierkegaard, a "public is neither a nation, nor a generation, nor a community, nor these particular men, for all these are only what they are through the concrete" (Kierkegaard 1962, 62). The opportunistic "qualification 'public,' [is] . . . . produced by the deceptive juggling of an age of reflection, which makes it appear flattering to the individual who in this way can arrogate to himself this monster, in comparison with which the concrete realities seem poor" (Kierkegaard 1962, 63). In other words, one can render out of the fictitious public the appearance of mass support for any view, belief, or cause.

Throughout *The Present Age*, Kierkegaard is concerned with the lack of physical presence in which individuals directly interact, because when the individual is "personally present" he has "to submit at once to applause or disproval for his decision" (Kierkegaard 1962, 62). While it is true that online media can be vehicles for dissent that would otherwise be squashed by oppressive powers,[5] such dissent often expresses itself as the absence of social ties. "Only when the sense of association in society is no longer strong enough to give life to concrete realities," writes Kierkegaard, "is the press able to create that abstraction 'the Public,' consisting of unreal individuals who never are and never can be united in an actual situation or organization" (Kierkegaard 1962, 60).

The reflective discourse that grows out of perceived "public" space creates the illusion of having taken a stance, having committed to a belief, and having acted upon it. In Western liberal societies, the feeling of moral entitlement, superiority, and empowerment often accompanies one's affirming, without action or commitment, views that are held by the "public." And nowhere do such feelings have more fertile soil than online.

The danger of online advocacy is that it can provide the semblance of moral significance. The more popular a site becomes, numerous the responses of likeminded individuals are, or accolades one social group confers onto another, the less one feels the need to act upon her online postings. Memberships in online groups can evoke a false sense of moral accomplishment via the transference of the site's achievements to oneself. This is reminiscent of the stoic Epictetus' illustration of the folly of vicarious merit:

> If a horse were to say proudly, "I am beautiful," one could put up with that. But when you say *proudly* [emphasis added], "I have a beautiful horse," remember that you are boasting about something good that belongs to the horse. What then, belongs to *you*? (Epictetus 2005, §6)

Similarly, when we boast about the moral achievements of the groups in which we are passive members, we boast about achievements of the organization and not our own.

Online organizations can additionally make one's haphazard postings seem morally significant. For example, since online disagreements rarely result in face-to-face exchanges, both parties feel freed of the restraint they would otherwise show. As a result, they tend to aggressively attack another's views, as opposed to arguing for their own. Such outpouring of pent-up anger, especially for the morally impotent person, can produce feelings of elation when one tricks oneself into believing that one has righteously and powerfully acted upon one's values. Put differently, one chooses to act in "bad faith," that is, one lies to oneself that one's actions are morally significant. The lie is often preferred over serious moral action, because serious moral action takes the time, effort, and face-to-face commitment that renders one vulnerable to dissenters.

The fundamental drive for the above tendencies seems to lie in the fact that social media, popular culture, news industrial complex, and academia encourage the self-congratulatory elation of holding views rather than acting upon them. For activists in name only, casual participation in online forums, tweeting, and joining organizations is merely a method of fetishizing what ought to be the deepest concern of one's very personhood: the free exercise of values through action. It is a type of fetishism, in that one intentionally produces feelings of pleasure and empowerment by reflecting (and not acting) on one's moral views. And it is likewise a form of moral materialism, in that one deems himself good by merely collecting memberships in organizations.

Kierkegaard is considered the father of existentialism, because he was interested in the seeming lack of intrinsic value of man's world, and hence the unavoidable responsibility of the individual to choose how she should act and develop herself as a person: "What I really need is to be clear about *what I must do,* not what I must know, except in the way knowledge must precede all action. . . . The crucial thing is to find a truth which is truth *for me,* to find *the idea for which I am willing to live and die*" (Kierkegaard 1996, 32). Insofar as the press, public, and internet can inhibit committed, purposeful, knowledgeable, and value-oriented action, they are enemies of personhood, in that personhood requires the capacity to knowingly act, and in doing so, purposefully develop a sense of self and value.

4. CLASSIFICATORY AND INDEXICAL THINKING IN THE INTERNET AGE

I argued in sections 2 and 3 that the leveling effect of the internet begins with idle curiosity and the poor epistemic habits that it fosters. In the last section, I argued that the internet poses an existential threat to those who carelessly engage it, because it can unwittingly function as a pleasure-inducing surrogate for committed action. In this section, I argue that there is a more primary way in which internet culture can predispose one to despair.

One's perception is that the world hinges upon an intricate network of practices, social expectations, and preestablished classifications. The ecology of practices and expectations surrounding classifications forms the basis of one's concept of an entity, as well as the perceived relations that it bears to other entities.[6] Classificatory structures thus generate an implicit and explicit worldview.

Web browsers, by contrast, function differently. Browsers identify and harvest millions of pieces of informational content. In a fraction of a second, they arbitrarily assign numbers to harvested content, which is then prioritized according to one's search parameters. Browsers thus permit nimble and ever-changing indices of relatable content based upon an individual user's interests, desires, and curiosity. In generating on-demand indices, the browser erodes prior classificatory structure.

The insidious component of the World Wide Web, as an indexing tool *par excellence*, is exactly what makes it such a powerful force for positive social change:[7] it circumvents traditional classificatory standards and creates myriad connections. This can appear to be a good thing, for the unquestioned acceptance of classificatory structures tends to negatively predispose one to alternative classifications—classifications that may avoid the bigoted or biased orientations of traditional structures.[8]

However, the organization of terms, as they are lumped together via web browsers, tends to cloud the specific ways in which they fit into a particular disciplinary structure and the normative rules that apply to it. This leads to misunderstanding and misuse of terms. For instance, instead of reviewing the entirety of a book, article, speech, etc., which often establishes the meaning of specific content and its role according to a particular community, author, or discipline, web browsing presents content in isolation from the context in which it is intended and imbedded.

The consequence of the internet indexing content according to the user's search and from multiple sources, local and global, is a muddled presentation of terms from divergent cultural and classificatory paradigms. The casual curiosity seeker lacks the skill or interest to understand the nuanced use, background, and implication of the terms that are searched, as these are shaped by the discipline, culture, or historical period in which they occur.

Without an understanding of the imbedded norms and values in which a community employs its classifications, one may very well become a rebel without an actionable world from which to rebel. This is a two-headed problem. First, in order for one to develop some degree of foresight regarding one's own impact, there must be some command of the classificatory structure in which a change is sought. For instance, if one wishes to create social change concerning the manner in which the Middle East treats women or homosexuals, one cannot efficiently or productively create change without first understanding how women and homosexuals are understood, the manner in which such understanding incites the behaviors that one wishes to address, and the social/political features that permit and empower its widespread occurrence.

The second part of the problem concerns Kierkegaard's conception of the public, that is, the false perception of association by allowing one to opportunistically create aggregates that have no social expression other than the perception of them. This may occur when the indexing tools of internet browsers bring together similar interests or redundant content into a false aggregate of staggering numbers. Such circumstances can create instances in which one fights for or against views that have no existence beyond that of perception. Hence, even if a "public" view concerns a position that one wishes to act upon, it may have no productive social outlet, because it concerns causes for nonexistent circumstances. When one becomes deluded in this manner by the "public," one's personhood is diminished, because one's capacity to generate a horizon of meaningful action becomes stunted.

5. EXISTENTIALISM WITH HEGELIAN UNDERTONES?

Hegel, unlike Kierkegaard, sought to detail the spirit of one's age and culture in order to understand the possibilities it engenders, which I, and others such as Fuller, perceive to be an important avenue of social epistemology.[9] The functional relations, economics, laws, and beliefs—including the norms and expectations of classification—all create a horizon of what a particular population believes to exist or to be possible or good. For instance, a fifth-century BC peasant from China could not participate in a Christian Church, become a quantum physicist, or start a liberal prodemocracy movement, because these social phenomena did not exist in his lifetime. Similarly, what one can do and how one chooses to act is, to some degree, what one's cultural upbringing has led one to believe is possible. In this regard, there is good reason to believe that each person's beliefs and actions are mediated by one's social embedded*ness*.

Kierkegaard passionately believed that a direct relation with God is the ideal moral standing. Social mediation is at best a distraction and, at worst, treads upon God's authority. There is a sense in which he ignores the social mediation between what humans are and choose to be, and the possibilities that are alive to them. In doing so, he and other anti-Hegelians limit the horizon of meaningful action. Kierkegaard's philosophy therefore cannot aid one's understanding of the outcome of one's actions in cultures with unfamiliar classificatory systems.

Perhaps more important, in regard to the existential threat posed by the internet, is the German idealist conception of personhood. Very broadly speaking, the German idealist tradition tends to view the ability to reason, and hence the capacity to understand one's own actions, as permitting personhood. For Hegel, the act of someone making choices according to a rational rule or value is an exercise of one's freedom. While Kierkegaard perceives the Hegelian tradition as relying too heavily on reason, rather than faith, emotion, or intuition, Hegel does not deny the relevance of nonrational impulses.[10] Rather, Hegel suggests that the rational ordering of desires permits one to rise above hedonism, for without the rational ordering of desire, one's personal choices are limited to what is immediately pleasurable.[11]

Additionally, existentialists tend to portray human beings as inherently intimate with their own freedom. Hegel makes a strong case to the contrary. He argues that one does not become fully aware of the extent of her freedom until one acts in the world in which her desires and choices are resisted. When one is resisted by the constraints of the social and natural world, one's individuality becomes more clearly differentiated from that which is not the self. In short, by overcoming resistance, we test ourselves and win for ourselves greater realization of our own personhood.

When one recognizes what one chooses and acts upon those choices, one begins the process of cultivating her agency, that is, a person who has some idea of their individuality and ability to act in the world. In this latter sense, we are only potential persons until we begin to act according to the values we choose for ourselves and thus begin the process of knowing ourselves.

If this very broad overview of Hegel's account of personhood is correct, there is a sense in which the internet does not merely make us prone to despair, but it also clouds one's understanding of the social world and the manner in which one can act in it. For instance, when the internet reinforces bad epistemic habits, offers misinformation, creates the false aggregation of content, and pacifies one's desire to act in the world, one loses sight of the horizon of meaningful possibilities. When such horizons become hindered, not only is the existentialist's project undermined, but one's very personhood becomes endangered.

In framing the problem of the internet in an existential light, we have uncovered an alternative to the traditional polarization found in the philosophy of technology: on the one side of this polarization is the belief that technology will permit us to transcend our current state of humanity (e.g., transhumanism), and on the other side is the belief that technology is destroying our humanity by predetermining the manner in which humans act within it.[12] By casting the problem or boon of technology in an existential light, we see that the primary imperative of personhood remains: a "person" must make a choice, take a committed stance, and act. Why must a person do such things? Because failing to do so, quite literally, makes one less of a person.

NOTES

1. My gratitude to John Lyne for extensive comments on an earlier version of this paper.
2. For more on this topic, see *Social Epistemology and Epistemic Agency: De-Centralizing Epistemic Agency*, edited by Patrick Reider from Rowman and Littlefield International, forthcoming.
3. Those who are held legally liable represent only a small portion of offenders.
4. Cf. Dreyfus, "Anonymity versus Commitment: The Dangers of Education on the Internet," 644.
5. For a good specific example, see Elizabeth Losh, "Hashtag Feminism and Twitter Activism in India," *Social Epistemology Review and Reply Collective* 3, no. 3 (2014): 11–22, http://wp.me/p1Bfg0-1Kx.
6. For more on this topic and its epistemic consequences, see *Wilfrid Sellars, Idealism and Realism: Understanding Psychological Nominalism*, edited by Patrick Reider from Bloomsbury Publishing, forthcoming.
7. F. Allan Hanson nicely summarizes this position in "From Classification to Indexing: How Automation Transforms the Way We Think." *Social Epistemology: A journal of Knowledge, Culture and Policy* 18, no. 4 (2004): 346.
8. Hanson summarizes this position well in ibid., 347–51.
9. Steve Fuller, "Social Epistemology: The Future of an Unfulfilled Promise," *Social Epistemology Review and Reply Collective* 3, no. 7 (2014): 33, http://wp.me/p1Bfg0-1wG.
10. Hegel does, however, clearly prefer the rational over the irrational.
11. Terry Pinkard, "Freedom and Social Categories in Hegel's Ethics," *Philosophy and Phenomenological Research* 47, no. 2 (1986): 209–32.
12. Jacques Ellul, "Ideas of Technology: The Technological Order," *New Perspectives Quarterly* 17, no. 3 (2000): 21–30.

Chapter 6

Existential Privacy and the Technological Situation of Boundary Regulation

Elize de Mul

The topic of "privacy" in a highly technological society has led to complicated discussions in the scientific field and popular media alike. Difficulties with a cohesive definition and the contradictory behavior of people—claiming to have high privacy concerns but at the same time exposing themselves online in various ways—have led to all kinds of bold statements. Public flag-bearer of transparency Mark Zuckerberg has more than once implied that privacy in our age has become obsolete. With the public embrace of the internet in the 90s and the cultural implications of a large-scale worldwide use of the digital network, a growing number of scholars have started to wonder about a possible ongoing erosion or even "death" of privacy. Hence, it seems that a philosophical analysis of the concept of privacy is necessary—if not critical—to secure its position in our contemporary society. In this chapter, I will show, on the one hand, using the idea of "excentric positionality" from philosopher and biologist Helmuth Plessner (1982), that privacy is an inherent, ontological characteristic of human being, and therefore cannot be declared "dead" as long as we remain human. On the other hand, I argue that new technologies bring about fundamental changes in our everyday experience of self and world which have consequences regarding the concept of privacy.

EXCENTRIC POSITIONALITY

In his magnum opus *The Levels of the Organic and Man* (1928), Helmuth Plessner works out a philosophical biology, distinguishing three different types of "positionality"—the relationship an organism has toward its own boundary. In contrast to his contemporary Martin Heidegger, Plessner does not approach this question in terms of temporality. Instead, he starts from a *spatial* perspective, thus grounding the modern subject in a materialistic manner and assigning importance to the bodily character of organisms.

All living organisms have a *boundary*. This is essentially different from having an *outline*, as is the case with an inanimate object like a couch. This object stops being somewhere—the floor, the air—and therefore can be said to have a contour. Having a boundary, on the other hand, implies that there *is* something on both sides of it, that there is an *inside* and an *outside*. An inside in this case should not be understood as the filling of the couch, but rather as a dynamic unity. This unity relates itself toward the environment that it is part of but separates from it at the same time. The "plant"—consisting of roots, nerves, cells, fruit, etc.—relates to its environment in the form of boundary traffic. A tree sucks up water with its roots and catches sunlight. In return, it also delivers things to the environment, like oxygen. Although the plant has a boundary—an inside and outside—it does not, however, have awareness. In other words, the plant *is* but *experiences* neither its body nor its boundary. In case of a plant there is a certain intentionality (growing toward light) and reactivity (leaking resin when being cut), but the plant itself is not aware of either of these things. Plessner states that a plant therefore has an *open positionality*—it is open toward the world without being aware of this openness.

Animals, the second category Plessner distinguishes, have a boundary, an in- and outside, too. An animal also shows intentionality and reactivity, but—in contrast to the plant—it has a more or less developed awareness of its surroundings. An animal experiences its relationship to the world, and therefore not only *is* a body, but *has* its body as well. It feels hunger, or cold. In the case of animals, Plessner speaks of a *centric positionality*; the animal experiences its environment from "within," from its "center."

With that, we come to the third and last category of organisms: human beings. Like plants and animals, human beings have a boundary, an in- and outside. Like an animal, human beings experience their body and the environment that they are part of from within. They can be hungry, or cold. Humans take a centric positionality in regard to their own being. There is, however, another characteristic that makes the animal *human*. Humans are not only bodies (like the plant), experience their body (like the animal), but they also experience their own experience. Human beings, in other words, not only are conscious of their environment, but also have *self*-consciousness. Self-consciousness is possible because human beings not only experience their selves from within, but also are able to experience their selves from the *outside*. In contrast to the other two life forms, humans not only show centric positionality, but *excentric positionality* toward their own being as well. They can take a "third-person perspective" upon their own existence. Human beings experience themselves simultaneously from a centric *and* an excentric perspective, that is, they are always in- *and* outside of themselves. Traditionally, the last-mentioned characteristic of human being is also known as "reflectivity."

INTERNAL, EXTERNAL, AND SHARED WORLD

While all organisms can be said to have an inside and an outside, this "double aspectivity" (as Plessner called it) appears differently in the various life forms. Because human beings stand outside of their own center, they experience their in- and outside in a fundamentally different way than animals. The animal, "trapped" in its own center, is "lived" by its own internal instincts and the outside world appears to it as a surrounding environment of stimuli. Because human beings can take a third-person perspective upon themselves, they can distance themselves from their own instincts and impulses and—to a certain extent—are able to control them. Also, by being excentrically able to "objectify" themselves, the surrounding environment becomes a meaningful *world* of interrelated objects, their own body being one of them. Within Plessner's model of excentric positionality, we can observe three "spheres" of human reality: an internal, an external, and a shared "world" (Plessner 1975, 343). These worlds should not be understood as separate spheres, referring to different ontological domains, but rather different perspectives which are encompassed within the individual. The external world [*Aussenwelt*] consists of things, plants, animals, and humans—everything that makes up the material world, *experienced as a significant whole* by human beings. Seen from a third-person perspective, we are part of this external world, together with all the other objects that surround us. Seen from a first-person perspective, we experience our internal world [*Innenwelt*], which consists of feelings, emotions, desires, dreams, etc.

The excentric position of the human allows for a third sphere of human reality: the shared world [*Mitwelt*]. The shared world is shaped by our very excentric positionality itself; it is a direct expression of it (Plessner 1975, 302). Where other animals also experience "shared being" (e.g., they are part of a herd or flock), it is typical of human experience that, from a second-person perspective, we are aware of and can reflect upon the perspective of the other. Although the shared world can be understood as a "social" sphere preeminently, it would even exist with only *one* human living in the world (Plessner 1975, 302–3). This is due to the given that a human being can experience himself from an excentric standpoint as "another." This outside perspective on the self brings about the mutual sphere. In it, we all come together as excentric beings (Plessner 1975, 203), aware of ourselves, aware of others. *This makes public self-awareness possible and inevitable*. We essentially are social and related beings.

Often, the shared world is explained as the sphere of culture (Plessner, 1975, 307–8) and technology. These phenomena can be seen as a direct result of the excentric positionality of human beings. If there is something that characterizes our species, it is probably the way in which we design, build, and use various artifacts, which give shape to our everyday world, ourselves, and the way we communicate with others. In comparison to other animals, the long process of evolution proves to have been thrifty with qualities when handing them out to us. But, although we are—to speak with Heidegger—being thrown (*geworfen*) into the world (Heidegger 1996a, 127), thanks to our excentric positionality, we are well aware of our shortcomings, and, thereby, able to do something about it. In this way, Plessner observes, human beings are *artificial by nature* (Plessner 1975, 309–21). We supplement ourselves bit by bit, by using technology—in the broadest sense of the word—without ever reaching a "closure." Technology is part of our excentric positionality, and can be seen as part of the shared world.

EXISTENTIAL PRIVACY

Although a dog will react to a stimulus from the environment, for example, his owner calling "Rover!," it does not have a comprehension of what it entails to be "Rover." Having an identity only is possible when *self*-consciousness exists (cf. Fichte 1982, 98). Identity in a person is possible because of the excentric position he can take upon his own being. Experiencing the self from the outside, instead of from within, leads to a consciousness about one's own boundaries, that is, one's in- and outside, one's place amid other things and within a shared world. We not only experience, like all animals (Plessner 1975, 294–95), our body from within [*Leib*], but at the same time we experience our body from the outside, as material body [*Körper*].

The excentric positionality of humans splits us, because it dissociates us from our united centric perception of ourselves. Our feet, because of our excentricity, are not only a lived instrument to tread upon the environment, but become an object between objects. This objectification of self can be alienating; it drives us away from our center. Seen from the outside we may appear as a dynamic unity (an organism), but this unity is not only a unity sensed from within, but also given to us as an external object. This discrepancy creates a "distance" between us and ourselves, and with that a longing to bridge the gap in our self-experience.

One "bridging tactic" is realizing ourselves. Because humans can experience themselves as a "unity" from an external perspective, it is possible to have a coherent "identity" and even to construct it actively, or at least to steer (away from) certain parts of being. We can, for example, "tell" others who we are by putting on certain clothes, using certain brands, reading certain books, or listening to a certain type of music (though this "self-realization" with the help of external objects never completely coincides with ourselves, *because* they are external). We "manage" our boundaries—the part of our self that others directly experience—to carry out a carefully constructed "I," that is, we present a "self" toward others in our everyday lives (cf. Goffman 1959).

Logically, having a boundary *and* being aware of this fact makes us existentially vulnerable. Where animals can be said to be fragile because of their boundary (they can get hurt, be seen, and are—in a sense—*trapped* within themselves because of their centric positionality), human beings are *vulnerable* in their existence because they are *aware* of the possible boundary complications. These "complications" can occur not only on a physical level (one is aware of the possible danger of getting into a car), but also on a more abstract level. Linking back to the concept of excentric positionality, we could state that we can get "damaged" within the three worlds of human reality; the external world (physical injury because of a car accident), the internal world (psychological damage because of self-castigation), and the shared world (social-cultural exclusion, discrimination).

In summary, a person manages his boundaries carefully on a physical, mental, and social level to prevent such injuries *and* to realize a "self."

In *The Environment and Social Behavior* (1975), Irwin Altman works out a model that focuses on the interpersonal regulation aspect of privacy. He conceives privacy as an interpersonal process of boundary control. An individual controls his boundaries and manages the level of "openness" toward all kinds of others that he deals with in the everyday shared world. It is important to notice that Altman's idea of privacy is *not* just one of covering the individual self, but also one of opening oneself up. It can be seen as an "optimizing model" that respects that any given situation and context asks for another desired privacy balance (an idea further developed by Nissenbaum 2009). Adding to Altman, I want to propose that in an age of a growing amount of "self-knowledge technology" (e.g., popular in the "quantified self" [QS] movement), privacy is no longer only a matter of *interpersonal* regulation but also of *interindividual* regulation, that is, control of an individual opening up and shielding himself from certain knowledge and experience of *himself*. This idea of privacy, as control over one's own boundaries and boundary control as a way to shield oneself from *and* open oneself up to others and oneself in various situations, is fruitful.

What I am proposing here is that privacy is an *existential situation* that is directly linked to the excentric positionality of human being. Having and managing (multiple) boundaries, and—most importantly—being aware of these boundaries, brings about certain "privacy situations" that deal with the relation between the individual and the world (internal world ⟵⟶ external world), between the individual and him- or herself (internal world ←→ shared world in terms of technology and culture that influence experience of the self), and between the others and the individual (shared world ←→ internal world).

Importantly, then, we should not understand technology just as an instrument that can achieve or undermine privacy. Technology is part of the *circumstances* within which human experience and "boundary control" take place. Technologically induced changes in these circumstances can bring about changes *in our very boundaries*, as well as in our *relation toward our boundaries*. Hence, technologies play an important mediating role, as they have the ability to objectify aspects of our life and transform them into objects in the external world. Moreover, to say that privacy is obsolete or dead is to say that we have changed so fundamentally in our being that the boundaries and boundary attitudes implied by being excentric (and centric at the same time) have fallen away. It is rather the case that our boundaries are not a fixed given, but take on new shape with each technological renewal that mediates our experience of world and self (Kockelkoren 2014, 183).

THE INFLUENCE OF TECHNOLOGY: THE PHOTO CAMERA

In the landmark essay "The Right to Privacy" (Warren and Brandeis 1890), it becomes clear that technological developments and (new kinds of) privacy situations are profoundly connected. The article is the direct result of a "privacy pain" typical of technological development of its era: the (improved)[1] photo camera. Having personally been confronted with the emotional distress caused by publication of photographs and accompanying gossip, Warren and Brandeis feel the need to describe this new type of distress and to provide juridical countermeasures to protect people from this technological conditioned "hurt." In their article, they define privacy as a right of the individual to be guarded against the unwanted distribution of information (with emphasis on publication in newspapers) about their private life. This private life includes emotions, sensory experiences, feelings, thoughts, actions, personal relations, writings, and sayings.

When we take a look at the new type of "hurt" brought about by the photo camera using the framework constructed above, it becomes clear that it has to do with a technologically induced change in both individual *boundaries*, as well as the *relation* of an individual toward his own boundaries. As argued above, we manage our boundaries in the process of self-realization, that is, disclosing our inner world and presenting an objectified "self." So too, the written word can be seen as an early technological extension of boundaries of the self; it gives way to self-presentation spatially removed from the "direct" body-object. In case of a written letter in which one explains his or her inner thoughts and feelings toward a friend, the control of self-presentation (chosen words and subjects) lies with the writer. In the case of gossip—someone writing about certain actions described to a third party—the control over one's self-presentation is at risk because of an intrusion on the technological extended boundary of self (in the form of written words) brought about by another.

An important difference between the "technology" of the written word and the photograph is that the latter brings about a *visual* boundary extension instead of a textual one. Rather than merely describing one's actions, emotions, etc. these can be *visually represented.* Physically, this means the possibility of a "doubling" of (physical) boundaries of the self. We can be *here* and *now* as lived bodies, and at the same time *there* and *then* as materialized body-(photo)objects. This technologically induced change in boundaries of the self brings about a change in our relationship toward our own boundaries as well. That is to say, this extension and materialization of self can give the individual more control over self-realization (e.g., staged "*selfies*"), but can at the same time lead to the experience of loss of control over self-realization (e.g., paparazzi photographically recording an action you don't want publicized). In both cases, an objectification of self-experience takes place. We can thus observe that the introduction of the photo camera brings about specific technologically induced changes in the internal, external, and the shared world of human experience. The photo camera makes the external world—and our body-objects in it—an object even more explicitly. The internal world in terms of "expressability" becomes something that can be highly staged or accidentally "caught on tape." And these digital images are increasingly playing an important role in our social lives.

These technologically induced changes in boundaries of self, including our relation toward them, means a shift in the phenomenon and concept of privacy as well. Although privacy introduced by Warren and Brandeis as "the right to be let alone" may be a too "one-sided" interpretation (focusing only on seclusion), their article implicitly lays the foundation for thinking about privacy as optimizing "boundary control." It develops the concern for personal control over public identity and the risk to such "identity management" due to new technologies. By showing how protection of person and property have undergone an evolution—from something mainly physical to including a more spiritual or intangible dimension (Warren and Brandeis 1890, 193)—they respect the way in which new technologies expand, duplicate, and complicate our boundaries. Subsequently, "right to privacy" means each individual retains power over the way and to what extent that individual is made "public."

THE INFLUENCE OF TECHNOLOGY: THE SELFIE

Although the phenomenon of taking a photo of oneself is not new, the large-scale popularity of the digital self-portrait in 2005 led to its being coined the "selfie." A selfie is a self-portrait, made with a digital camera (often on a smartphone), capturing the face and parts of the body by stretching the arm or using reflective surfaces like a mirror. Because many smartphones are connected to the internet, selfies may be easily published online. In the Oxford English Dictionary, the publication of the digital self-portrait on a social network site is even a criterion for the "selfie."

The phenomenon has led to discussion in the public domain as well in the academic world about the apparent lack of privacy awareness of "selfsters." Indeed, it is not uncommon to see selfies in which the subject is on a toilet, half naked, in distress, ill, or a private situation like a deathbed or birth. For skeptics, these types of selfies form a perfect example of the "privacy paradox": expressing privacy concerns, yet behaving seemingly contradictory. However, looking at the selfie phenomenon from classic "normative" conceptions of what ought to be private does not respect changes brought about by new technologies. Hence, we should approach the selfie, and the technological mediation making the phenomenon possible, in terms of the above noted right to privacy (Warren and Brandeis) and in terms of our boundaries and relation toward our boundaries (Plessner).

Consider how analysis of the word *selfie* shows that it is mainly used as a noun. Yet, when we look at the process of making selfies—which we may call "to selfie"—the active constructing character of the phenomenon becomes clear: *a self is being made*. Though during this process the control lies to a certain extent with the subject himself, it is not uncommon during a "selfie session" to take multiple photos, thereby sampling different "selves" for selection to be published for others to see.

However, in regard to the technological situation of boundary regulation, one should not forget that one cannot completely control the effective history of these identity constructs. For example, others may graft these images in contexts not foreseen by the "selfster" through various means of technological mediation, that is, Facebook, Twitter, or Instagram. By not looking into such (de)constructive aspects, brought about by new technological possibilities, certain dimensions of contemporary reflective being, identity construction, and privacy are easily overlooked or misinterpreted.

PRIVACY RESEARCH AND NEW TECHNOLOGIES

Privacy has been studied for more than 100 years in practically every realm of social science. Because it is difficult to pin down what exactly "privacy" entails, that is, given the wide range in which the concept may be developed across disciplines with different applications in various domains, often "privacy" is used as an umbrella term, or is coupled with other concepts (e.g., security, autonomy, freedom, integrity). Similarly, the emotional character of public and media discussions does not necessarily make use of the term any clearer. A consequence is that though we may recognize that the application of (new) technologies in new domains brings forth new types of "privacy issues," we require different conceptual tools and nuance to understand these issues properly. Without a different conceptualization, often we find a narrow use of "privacy" at work as the mechanism of keeping some information hidden and some in view, or the understanding of privacy as some type of "value" that logically can be more or less valuable than other values (like security). Moreover, such a narrow use of "privacy" does not do justice to the complexity of being situated in a highly technologically mediated social world.

Privacy understood as an existential aspect makes room for the insight that the phenomenon of privacy is fundamentally altered with technologically induced changes in our boundaries and our relation toward our boundaries. Hence, existential privacy may be seen as integral for research projects concerned with social epistemology generally or to raise public self-awareness specifically, and to do so one would start with describing and analyzing our contemporary excentric positionality regarding the constructions from our highly technologically mediated social world, in order to be able to distinguish these new "types" of privacy. As it currently stands, the debate on privacy seems to skip over an examination of our contemporary technologically influenced being-in-the-world, thus missing the parallel changes in the meaning of privacy.

As I have proposed above, this can lead to misinterpretations of certain phenomena, as well as to bold statements that—from an existential perspective—do not make sense. The assumption that, in our contemporary society, privacy is somehow "dead" overlooks new technologically induced existential possibilities and the accompanying technological situation of boundary regulation.

CONCLUSION

As we have seen, new technologies—as part of our very being as humans—alter not only our actions in the world, but the world itself and our place and identity in it. As a consequence, fundamental concepts like "identity" and "privacy" gather new meaning within these new contexts. To protect the position of privacy and to shed light on the new ways in which we *are* and experience ourselves and the world, it is important to go to the roots of what (contemporary) privacy is. I have argued that—understanding human being's excentric positionality—privacy can be understood as an existential aspect of being human. Technologies introduce changes in our boundaries as well as our relations toward these boundaries.

Consequently, the concept of privacy, as I explained above in terms of boundary control mechanisms, changes alongside with new technologies and technological mediations. I have argued that current discussions seldom recognize this, thus failing to see new boundaries and the connected new "boundary issues" that are brought about by new technologies and the way we use them. Hence, further philosophical analyses regarding boundary regulations are warranted, including but not limited to those developments of, or projects influenced by, Plessner's philosophical anthropology. At stake is the concern that privacy may be deemphasized in favor of better defined concepts (like security) in social epistemology and scientific fields as well as in our everyday lives.

NOTE

1. By the time Warren and Brandeis wrote their article, cameras were small enough to be handheld and were able to capture an image at a much higher speed than their ancestors, which required the subject of the portrait to hold still longer.

Chapter 7

Critical Media

Media Archeology as Critical Theory

Stephen M. Bourque

The aim of this chapter will be to explore the intersection between media archeology and critical theory and to develop a sense of "critical media." Media archeology is concerned with history, specifically the changing notions of media and the changing interactions that subjects have with it. That is, media archeology may be understood as the application of critical theory to the effects of technological mediation, both historically and socially. These changing notions and interactions may be further understood as constituting the material of social epistemology.

Just as part of the aim of critique, from the work of Immanuel Kant to Michel Foucault, has been to develop a philosophy reflexive regarding its present time, this chapter invokes critical theory regarding technological mediation toward increasing public self-awareness. That is, by developing the relation between critical theory and media archeology, this chapter advocates the methodology of critical theory to engage the interdisciplinary nature of the material of social epistemology. Interdisciplinary, here, is intended to highlight the multiplicity of discourses, ways of practicing, and methods of inquiry involving both the humanities and sciences. Hence, this chapter offers an understanding of critical media from which public self-awareness regarding technological mediation may emerge.

Though the inherent vagueness of media archeology, which is in part due to its multiplicity of techniques, has allowed for the continuing exploration of multiple relations between things, people, perspectives, and methodologies, the term "media archeology" is infamously difficult to define. Beyond the etymology of the term, then, which suggests the study of the origin of media and the various historically, culturally, politically, and technologically grounded relations which mediate everyday practice and social knowledge, that is, social epistemology, the following blog post from media archeologist Jussi Parikka helps situate the term.

> Media archaeology has succeeded in establishing itself as a heterogeneous set of theories and methods that investigate media history through its alternative roots, its forgotten paths, and neglected ideas and machines that still are useful when reflecting the supposed newness of digital culture. (Parikka 2010)

Further, "Media archeology is decisively non-linear," meaning it attempts to study all types of mediation such that it must "insist both on the material nature of its enterprise . . . and that the work of assembling temporal mediations takes place in an increasingly varied and distributed network of institutions, practices and technological platforms" this, Parikka stresses, involves even the examination of "how technology is the framework for temporality for us" (Parikka 2010). In this way, media archeology as critical theory may also be applied in regard to social epistemology.

I. POWER, SUBJUGATION, AND CRITIQUE: FOUCAULT AND CRITICAL THEORY

The work of Michel Foucault helps locate the idea of critique in relation to French critical theory or, more specifically, to power structures in general. Though Foucault's method may have changed over the course of his career, he maintained a conception of self-reflecting critique and an examination of the constitution of power regarding bodies and discourse. Foucault's idea of critique most heavily relies on the analysis of power relations and their importance in micro-political or macro-political action. First and foremost, according to Foucault, power relations are inescapable: we are always bound to power's grasp, through its dispersal and executed action in a given discourse or body (as well as intersubjectively).

In *The Archeology of Knowledge* (2002), written in the middle of his career, Foucault attempted to define, for critics and colleagues alike, the motivation and the methodology that fueled his previous published works (cf. Foucault 1994, 2001). Thus *The Archeology of Knowledge* holds significant importance regarding Foucault's methodology and the development of his idea of critique. According to Foucault's characterization, "My aim is most decidedly not to use the categories of cultural totalities in order to impose on history, despite itself, the forms of structural analysis . . . but are intended to question theologies and totalizations" (Foucault 2002, 17). He continues (in reflecting on previous works):

> The studies of madness and the beginnings of psychology, of illness and the beginnings of a clinical medicine, of the sciences of life, language, and economics were attempts that were carried out, to some extent, in the dark: but they gradually became clear, not only because little by little their method became more precise, but also because they discovered—in this debate on humanism and anthropology—the point of its historical possibility. . . . In short, this book, like those that preceded it, does not belong . . . to the debate on structure, it belongs to that field in which the questions of the human being, consciousness, origin, and the subject emerge, intersect, and separate off. (Foucault 2002, 17–18)

Specifically, the critical aspect of this reflection, which is his main concern, may be characterized in the following questions: How is the identity of being human structured in different periods of history? How does power leave its mark on the subject? And, how does the modern conception of power constitute bodies and discourse? After *The Archeology of Knowledge* Foucault began to address his place in critical theory more generally, for example, in his turn toward genealogy, power, and bodies.

Foucault considered the following to be the objective of his 1974–1975 lectures at the Collège de France (Foucault 2004a) course:

> What I would like to study is the emergence of the power of normalization, the way in which it has been formed, the way in which it has established itself without ever resting on a single institution but by establishing interactions between different institutions, and the way in which it has extended its sovereignty in our society. (Foucault 2004a, 26)

Foucault's move toward the powers of normalization in this lecture and its published counterpart, *Discipline and Punish* (Foucault 1995), utilizes an example of the way in which power subjects identities and bodies to roles that it plays in the development of the relations between institutions, individuals, and society as a whole. These relations are also integral to the study of social epistemology.

These studies located after *The Archeology of Knowledge* indicate a shift of emphasis to the role that power and genealogy play in the context of modernity:

> Genealogy is, then, a sort of attempt to desubjugate historical knowledges, to set them free, or in other words to enable them to oppose and struggle against the coercion of a unitary, form, and scientific theoretical discourse. The project of these disorderly and tattered genealogies is to reactivate local knowledges . . . against scientific hierarchicalization of knowledge and its power-effects. (Foucault 2003, 10)

While seeming not to reflect on the modern situation, genealogy allowed Foucault the opportunity to show the development of power in relation to social institutions, thereby shedding light on modernity. Whereas the critique of modernity is immanent throughout Foucault's genealogical works, his notion of "power" is the key to Foucauldian critique. In fact, he argued that such examinations are what "alone allow us to see the dividing lines in the confrontations and struggles that functional arrangements or systematic organizations are designed to mask" (Foucault 2003, 7). Hence, Foucault's idea of critique examines the structuring of subjects and their immanent place in power's grip. Foucault's form of critique, then, relies on the immanence of the subject caught in power relations as well as the ability for the subject to resist, revolt, and be subjugated and oppressed by such conditions.

II. MULTIPLICITY OF CRITIQUE: THEORY, DISCOURSE, AND MEDIA ARCHEOLOGY

Media archeology's concern with understanding present phenomena in a critical vein gives its discourse the power of potential critique. Media archeology, at times, expresses a relationship to the possibilities of modernity and modern experience. In order to do justice to media archeology, the implementation of categorization techniques will be used in order to organize some of the different approaches and methodologies that different media archeologists use in their discourse. The following categories are intended to be liminal and fluid, rather than distinct and exclusive. The three categories are: (A) Theoretical Media Archeology; (B) Archival Media Archeology; and (C) Median Media Archeology. Recalling Parikka's insistence noted above, whereas theoretical media archeology looks at mediations that take "place in an increasingly varied and distributed network of institutions, practices and technological platforms," archival media archeology's focus is "on the material nature of [media archeology's] enterprise" (Parikka 2010).

A. Theoretical Media Archeology: Critical Media I

In regard to theoretical media archeology, Wendy Hui Kyong Chun, Matthew Fuller, and Andrew Goffey are prime examples. This section considers Fuller and Goffey's Felix Guattari-inspired *Evil Media* (2012). The idea of critique which *Evil Media* suggests may be located between pure theoretical conjecture and an archival approach to objects. It traces the boundaries of these objects as it examines the in-between area of media, ethics, theory, and linguistic strategy. *Evil Media* exemplifies theoretical media archeology, then, in its employment of a provocative style in relating to the objects it studies.

The specific textual argument of *Evil Media* is that media and the concept of mediation in general form a unique aspect of our interaction between objects and concepts (cf. Adorno 1973). Hence, this text is an important continuation of the idea of critique that moves between disciplines, opens up innovative views of modernity, and forms another tool in the multiple methodologies of not only media archeology but also critical thinking in general.

Evil Media is an innovative work that pushes the boundaries beyond criticism, media studies, and theoretical argumentation. Written in "stratagems," this work is indebted to the literary form of the fragment and aphorism. The importance of this style, reminiscent of Theodor Adorno's work, is imbedded in the concept of mediation for which the author argues as well as the role of interpretation in understanding the technologies that mediate our everyday existence:

> The technology itself is inseparable from the practices of which it is a part, and it is in the way that aesthetic qualities conjoin with organizational practices, roles to play, appearances to manage, and so on, that technology perhaps accomplishes its most powerful effects. (Fuller and Goffey 2012, 10)

Coupled with acknowledging the role that technology plays in mediation, this term is further defined by the "grayness" that the experience of media (as mediation) plays in our banal, everyday tasks. The object of the stratagem as an organizing function of *Evil Media* is a tool in expressing the general thesis of the often unexamined nature of "gray" media and the way that they, much like the stratagem, rupture, become confused and convoluted, and structure our interaction with ourselves and society. Consequently, exploring this "grayness":

> Gives rise to an experience of the vague, to fuzzy experience. To escape the bland feelings that blend into the background like a steam into clouds, a little clarity, definition, or even friction is required. (Fuller and Goffey 2012, 11)

This friction, valued so highly by the text's authors, is what the stratagem forcefully exposes its reader to. This is a work of "background mediation": mediation that we take for granted as to its social and political importance in the tasks and organization of our everyday life. For example, "A software engineer cannot avoid making assumptions about how an application or tool will be used, and such assumptions are ripe for exploitation in more ways and more sense than one" (Fuller and Goffey 2012, 8). *Evil Media* seeks to further the importance of media as a concept and moreover to promote an idea of cultural critique that articulates its performance in an "event-ness": a moment of rupture and reflection.

Evil Media, in addition to its iconoclastic methods of exploring media and mediation in general, is also unique in its inquiry of research and interdisciplinary blending of objects, concepts, language, and academic imagination. The text implicates itself in the archeological method through its intense research of media. However, much of *Evil Media* takes media as its object, yet consistently extends beyond the simple object toward ways in which these critical media aphorisms examine the idea of possibility as a dynamic that can open up knowledge and praxis to a more critical examination:

> A basic scenario sketched out across the range of stratagems we have explored here is one in which some technique, technology, tool, device, practice, chemical, concept or other such thing operates to shape and configure the possibilities of a situation, influencing the way in which it can change, the dynamics operative within it. (Fuller and Goffey 2012, 125)

The hope of the text is to implement these inquiries in order to shape the possible encounters that a person has with media. Ultimately, the idea of the stratagem and the exploration of these objects on a theoretical, cultural, political as well as linguistic level are in order to activate a form of resistance given to one who lets one's imagination be ruptured by these concrete yet theoretical studies. *Evil Media* offers a different form of critique that captures intense research in an archive coupled with the ability to offer a form of discourse that oscillates between concept, object, and reflectivity all the while realizing the inescapable nature of all these things, which makes the text explicitly conscious of its immanence in the interaction that an individual has between media and itself.

B. Archival Media Archeology: Critical Media II

On the material side of theory in the practice of media archeology as a discourse is the act of retrieving information from the archive and presenting it in a text. While this archeological method is not a recent invention, the inspiration for this form of discourse is indebted to the archeologists and philosophers of the twentieth century. Describing the archive, Foucault states,

> The never completed, never wholly achieved uncovering of the archive forms the general horizon to which the description of discursive formations, the analysis of positivities, the mapping of the enunciative field belong. (Foucault 2002, 148)

According to Foucault, the archival research project has a never-ending job of presentation and development. This analysis is important in the research that it performs in presenting new perspectives and retrievals of information toward the re-representation of the modern. Drawing from this conception of Foucault and the project of archive researching, archival media archeology spends its time within the labyrinth of documents, media, and information in order to present new ways of relating these findings to antiteleological historical formations as well as reconceptualizations of modernity. While this research is at times problematic it is nonetheless important work in continuing to assess modernity in its relation to history. Archival media archeology is best represented by the work of Daniel Rosenberg, Anthony Grafton, and Siegfried Zielinski. The seemingly noncritical edges of this work will be sharpened through its implementation in modern media archeological discourse and interaction with critical theory.

In Daniel Rosenberg and Anthony Grafton's *Cartographies of Time: A History of the Timeline* (Rosenberg and Grafton 2012), the development of the timeline is riddled with new documents presented in beautiful graphic detail. The work seeks to examine the history of cartography and its relation to chronology. However, it makes no implicit argument for a progressive history. Rather, it traces the differences in the documentation of time that have occurred from early Greek notions of the timetable to the modern phenomenon of the timeline. While the basis of this text is mostly the presentation of materials dug up in archival research and an amalgam of facts attached to these representations of chronology, the authors state,

> The timeline seems among the most inescapable metaphors we have. And yet, in its modern form, with a single axis and a regular, measured distribution of dates, it is a relatively recent invention. Understood in this strict sense, the timeline is not even 250 years old. How this could be possible, what alternatives existed before, and what competing possibilities for representing historical chronology are still with us, is the subject of this book. (Rosenberg and Grafton 2012, 4)

Given the author's acknowledgment that their presentation of materials is without teleology raises two unique questions: Is this simple presentation of materials more in depth than the authors seem to acknowledge? What is implicit in the presentation of antinarrative historical documents that adheres to an idea of critique? *Cartographies of Time* is an excellent representation of an archival media archeological text that is performing a basic task of presenting archival research. However, the presentation of this text executes a more or less relatively straightforward research project. If, as thinkers, we are to take seriously Benjamin's idea of the story as disruptive in its own right, the presentation of materials and their context already shows materials in a certain way and cannot remove itself from creating a narrative and history in which the ideas, materials, and people are simultaneously connected.

In contrast to *Cartographies of Time* stands a work that reflects on its close proximity to its archival research and its own story-telling function. The heavily researched work by Siegfried Zielinski titled *Deep Time of the Media* (Zielinski 2006), developed in the depths of the archive, is a work that presents a unique way of integrating research with narrativity and stands in stark contrast to the presentation of the materials in *Cartographies of Time*. Zielinski openly acknowledges that when doing away with the notion of progress in history, one now must construct a disjunctive narrative:

> If the interface of my method and the following *story* are positioned correctly, then the exposed surfaces of my *cuts* should *reveal great diversity* [emphasis added], which either has been lost because of the genealogical way of looking at things or was ignored by this view. (Zielinski 2006, 7)

The story's hope is that the cuts (i.e., the decisions it makes about which archeological material to present) made by the reconstruction of a narrative, lead the way to "great diversity" or to a conception of history that embraces heterogeneity. These paths are stories designed to circumvent, intersect, and disperse the grand narratives of media history and history in general:

> Possibly, one will discover fractures or turning points in historical master plans that provide useful ideas for navigating the labyrinth of what is currently firmly established. In the longer term, the body of individual anarcheological studies should form a *variantology* of the media. (Zielinski 2006, 7)

The idea is to generate contradiction and ambiguity, but more importantly to acknowledge history's imagistic construction and the appearance of images connected in a story told by a historian. Zielinski's work in *Deep Time* acknowledges this construction and even embraces the idea of the archeological story that cannot be removed from the historian's work. This embrace is met in order to produce heterogeneity:

> Magical, scientific, and technical praxis do not follow in chronological sequence for anarcheology; on the contrary, they combine at particular moments in time, collide with each other, provoke one another, and, in this way, maintain tension and movement within developing processes. (Zielinski 2006, 258)

The construction of a story (whether following chronology or not) creates an amalgam of moments that collide and in turn produce movement. Contradiction in history, when combined together in a narrative, gives the opportunity to maintain irreconcilable dimensions of history and open new possibilities for critique that allows one to position him- or himself to be more responsive to modernity's present contradictions.

Zielinski is well aware of the archivist's inescapable narrative construction when presenting materials found in the archive. *Deep Time of the Media* embraces the immanence inherent in presenting any type of historical work in a given medium and the historical work's nonremovable assumptions about the narrativity of history. Zielinski's work is not simply a counternarrative, it embraces the structure of the story and through its own story reflects on the importance that this form of discourse has for critique in general. For Zielinski, the story is an *image* of a reality located within a heterogeneous multiplicity of competing reality-conceptions, all occurring together at once; past and present locked in an immanent struggle of stories based on facts, interpretations, narrativity, and ideas.

C. Median Media Archeology: Critical Media III

Located in between theoretical media archeology and archival media archeology is the middle place of the discourse, named for our purposes "median media archeology." This type of media archeology combines intense research and new archival objects, all the while consistently reflecting back upon larger theoretical and cultural implications of the study of these objects. The inherent theoretical consequences of this concern open up larger philosophical questions that are implicit within the texts. Median media archeology typically begins with an object, moves into the historical analysis and implications of that object to theoretical consequences and opportunities, and back to the object again. The authors who characterize this type of work are Jonathan Sterne, Lisa Gitelman, and Erkki Huhtamo.

Jonathan Sterne's work "MP3: The Meaning of a Format" (Sterne 2012) is another text concerned with the nature of media as mediation: "Mediation is not necessarily intercession, filtering, or representation. Another sense of mediation describes a form of nonlinear, relational causality, a movement from one set of relations to another" (Sterne 2012, 9). His study is an analysis of the format and the various ways that different forces come together and separate apart in the formation of standardizing practices. Standardization of a format entails all types of development in technology, cultural practices of exploitation and marketing, as well as individual innovation and bureaucratic intervention.

While tracing this phenomenon through the development of one technological object, Sterne makes clear that his objective is "a story of how a set of problems in capitalism taught us to think about hearing and communication more broadly" (Sterne 2012, 30). Thus, Sterne's work explores the theoretical questions of modernity in relation to hearing and communication while describing the development of a specific set of practices and technological objects. Moreover, the history of these forces is simultaneously a history of the MP3 format. While examining the history of compression of early audio formats, Sterne still finds the time to examine the development of compression as a means of administrative control:

> The lack of price competition enabled more regimented administration and planning within corporations, and also forced them to look for other ways to increase their value and profitability—either through finding cost-saving measures in the production process, through other innovations, by expanding their operations, or by acquiring other corporations. More than ever before, corporate capitalism became a matter of administration, and administration itself became one of the central dimensions of both economy and culture in twentieth-century life. (Sterne 2012, 42)

Paragraphs such as these are riddled throughout "MP3," and add to the depth of the study. These moments also maintain a constant reflection between the object of study, history, and larger theoretical consequences.

In this vein, Sterne's work presents its readers with a changing idea of critique developed from his indebted history to critical theory. Building on and furthering the project of Adorno, Sterne's immanent critique comprehends that it must take the object as its starting point and move between theory and the analysis of the object, in order to understand it more fully. This is not making the assumption that the phenomenon becomes actual, but accounts for the way in which discourse and critique must make into objects things it seeks to release from objectification. Further, Sterne acknowledges the challenge that is inherent in objectification in discourse.

His challenge for critique is its ability to "assess our abstractions" (Sterne 2012, 243), that is, reflect on the normative terms that we take for granted in discourse. These allow one to find in an object its micro-political as well as its macro-political situation. Median media archeology, in practice, maintains this distinction that Sterne is advocating for. It practices the oscillation between the particular and universal and the tension of the object in the midst of society. It challenges its examination by realizing the different institutional constructions that an object may be swept into through its implementation in society. These factors further the critique begun by Adorno and Foucault by challenging the critic to examine a practical, material object all the while consistently reflecting back on the object's place in the structure and how these structures effect its material and ideational possibilities.

III. CONCLUSION

This chapter has sketched an outline, reviewing some foundational literature, of critical media studies. Though still nascent, critical media studies may be characterized in terms of its affinity to critical theory, whether as theoretical media archeology, archival media archeology, or median media archeology, "critical media studies" refers to a heterogeneous set of theories and methods for critically investigating media. In this way, critical media studies should be efficacious for social epistemology in critically examining the social effects of technological mediation, for example, regarding knowledge and the various interactions afforded by social media.

Critical media studies aims at the acknowledgment of mediation overall and the conceptual importance of the "gray" area between our concepts and objects. This inescapable attribute of mediation is an important thread between media archeology and critical theory. Between Foucault, Fuller, Sterne, and Zielinski are approaches and methodologies that address the idea of critique, mediation, power, and the effects of culture more generally regarding subjectivity, discourse, and everyday living. Hence, critical media studies should have value for inquiries in social epistemology, such as "How should the pursuit of knowledge be organized?"

Chapter 8

Speculative Ethics and Anticipatory Governance of Emerging Technology

A Case for "Un-disciplined" Philosophy of Technology

William Davis

Some ideas, like the following from David Collingridge's "dilemma of control," present claims so intractable, they actually halt intellectual traffic.

> The social consequences of a technology cannot be predicted early in the life of the technology. By the time undesirable consequences are discovered, however, the technology is often so much part of the whole economics and social fabric that its control is extremely difficult. This is the dilemma of control. When change is easy, the need for it cannot be foreseen; when the need for change is apparent, change has become expensive, difficult and time consuming. (Collingridge 1980, 11)

For philosophers of technology interested in providing prescriptive analyses and judgments of emerging technologies, the above example directs us to an impasse. Technologies are far simpler to govern in the early phases of development and implementation. However, their social, economic, and political impacts resist early identification and tend to be realized only after technologies have permeated societies and cultures. Unfortunately, attempting to control and regulate technologies after they enter the marketplace becomes so laborious and economically arduous that efforts to do so strain resources (Johnston 1984). Where and when should philosophers of technology enter such deliberations? Speculative ethics and anticipatory governance offer a tentative direction: explore and evaluate, in advance, the values, politics, and social goals that societies desire their technologies to promote. They invite us to take an active role in shaping the technologies that will inhabit, and significantly shape, our futures. How can we best organize and implement such inquiry? By leveraging users, lay publics, and experts (academic or otherwise), an "un-disciplined" philosophy of technology (UPoT)[1] offers a step in the right direction.

As David Guston (2014) notes, Science and Technology Studies (STS) scholars—of which philosophers of technology comprise one group—champion approaches to governance of scientific and technological developments that emphasize nonlinear and co-constructive flexibility (Ozdemir et al. 2009), avoid top-down policy pronouncements (Groves et al. 2010), encourage public participation (Quay 2010), and promote responsible innovation (Owen et al. 2012). Philosophical inquiry into "the good life" in this century demands that the sciences and technologies purportedly improving our lives and our world match the values that our societies wish to promote and/or explain the changes that should be made (Michelfelder 2011). Such demands, however, require philosophers of technology to eschew "end of the day assessments" that come after the technology has entered societies. We must offer speculative judgments about the kinds of human-technology-world relationships that will permit democratic and participatory intervention once emerging technologies leave design spaces and enter into the complex social, economic, and political frameworks of our daily lives.

This chapter reviews recent debates about speculative ethics regarding emerging technologies, but it also offers a proposal for making philosophical analysis of such technologies integral to anticipatory governance discussions. Of particular concern is how to make philosophical discussion—including the ethical, political, and social impacts—of emerging technologies available and accessible to lay audiences. I offer paths forward for "un-disciplined" philosophers of technology performing speculative ethics and engaging in anticipatory governance regarding, for instance, nanotechnologies (NTs) and information and communication technologies (ICTs). I argue, *pace* Nordmann (2007), Keiper (2007), and Nordmann and Rip (2009), that speculative ethics of emerging technologies demands attention not only from ethicists (Roache 2008) and academic philosophers of technology (Michelfelder 2011), but also from the publics, institutions, end users, and unintended patrons that will be affected. Specifically, I focus on what a normative and UPoT can contribute to speculative ethics and anticipatory governance[2] to aid in developing, organizing, and critiquing our normative visions regarding emerging technologies and the kinds of societies we will construct in this process. A populace sufficiently informed, motivated, and enabled to enter discussions about the values promoted by our chosen technologies will not resolve Collingridge's dilemma. It would, however, provide a basis for broader, more accessible debates about the types of societies that we wish to live in, and the technologies that may propel us toward, or hinder us from, reaching those states.

UN-DISCIPLINED PHILOSOPHY OF TECHNOLOGY AND ANTICIPATORY GOVERNANCE

The last thirty years have seen a shift to micro studies and practice-oriented philosophy of technology (examining "technologies," not Heidegger [2008], Ellul [1964], and Marcuse's [1991] "Technology").[3] These useful engagements incorporate constructivist accounts of technological development and propose that philosophers of technology should engage engineers and designers of developing technologies (Ihde 1990; Verbeek 2008, 2011b, 2012; Dorrestijn 2012). By restricting their study to specific technologies, however, they miss the opportunity to provide speculative accounts of possible futures enabled and made possible by the still-emerging applications of NTs and ICTs, as well as current technologies related to genomic engineering (Baltimore et al. 2015). UPoT seeks to include more voices and engage more audiences than are typically found in academic journals. They speak directly to their readers because their topics, if not their positions, bind us together, as individual users of the technologies simultaneously part of a larger system.[4] Un-disciplined philosophers of technology, like Kevin Kelly (2010), Ray Kurzweil (2005), Jaron Lanier (2011), and Evgeni Morozov (2012, 2013a), have made inroads into popular press outlets by communicating straightforwardly to their readers, without the weight (and, consequentially, without the support) of copious references, footnotes, and technical jargon. If we wish for speculative ethics and anticipatory governance of emerging technologies to become publicly traded discussion topics, then we need to change how we communicate about them: these debates must become habit.[5] UPoT contributes to organizing and implementing visions of how we want to develop emerging technologies, including incorporating values into the technological designs themselves.

Without proposing a return to analyses of capital "T" Technology (as opposed to the current move of micro analyses), I argue that a UPoT can reinvigorate broad discussions of technologies' impact on the human condition. By incorporating lay publics into discussions and determinations about the values—moral or otherwise—that technologies should promote, UPoT can offer speculative ideas about the kinds of future societies that these technologies would enable. For anticipatory governance to be broadly participatory, lay publics must realize their potential to influence developing technologies and, thus, societies, political systems, ecologies, etc.

The shift we need to make, emphasized by academic philosophers of technology like Feenberg (2002a) and "un-disciplined" philosophers of technology like Kelly, Lanier, and Morozov, pushes us to examine the values embedded in our technologies and ask if these are the values we wish to promote. One central issue, however, remains: Where will such discussions and debates take place, and with whom? Two current options exist. The first consists of university classrooms, and centers for research, like Arizona State University's Center for Nanotechnology in Society (CNS). An informed lay public requires accessible communication of the ethical, legal, and social implications/applications of technologies. University classrooms and research centers, as well as informative opinion pieces like those from think tanks (cf. Wittes and Chong 2014), serve as ideal incubators for such education, but they also perpetuate top-down styles of governance that only reach academic or elite audiences.

The second arena, like focus groups and computer-based interactive spaces, present emerging technological developments and invite reader feedback. The *MIT Technology Review* and *Slate's Future Tense*[6] both offer commentary on developing technologies and provide spaces for lay publics to respond. Simply offering a comments section, however, is insufficient for actually making spaces for lay publics to interact with developers of emerging technologies and scholars working in anticipatory governance. We need to reimagine the roles and interactions of lay publics with technology makers, policy analysts, and activists if we wish for anticipatory governance of emerging technologies to be more than an academic affair. We must further explore notions of "lay publics," and consider who should serve as interlocutors for speculative ethics and anticipatory governance of emerging technologies. The first step involves presenting descriptions and normative implications of emerging technologies to lay publics. The second—and most difficult—step requires the creation of truly interactive arenas where the publics and experts can exchange ideas and make recommendations for future developments.

ANTICIPATORY GOVERNANCE

As Steve Fuller notes, the

> omnipresence of anticipatory governance is felt in the proliferation of focus groups, consensus conferences, Internet surveys, and Wiki and other interactive media—all of which, again intentionally or not, serve to cast doubts on the representativeness of classic democratic institutions like legislatures and elections. (Fuller 2010, 533)

Supplying feedback, solicited or not, to elected officials regarding the types of policy for emerging technologies, does not amount to the type of engagement that permits publics to actively participate in decision-making processes. Furthermore, as Guston (2014) explains, lay publics do not possess sufficient knowledge of emerging technologies like NT to make positive contributions to policy decisions.[7]

The challenge will be both to inform lay publics and to provide them with spaces to have their voices heard. Un-disciplined philosophers of technology are not restricted by the same constraints as academic philosophers. They can disregard philosophy's professional practices—styles of writing/presenting, as well as interactions with audiences outside of single disciplines, perhaps even those not affiliated with any particular discipline—and thus transcend a subdiscipline of philosophy. In doing so, they provide their audiences with the means to examine the human-technology relationships that shape increasing aspects of our existence.

UPoT must present technological development as co-constructed—context-dependent and coproduced (dependent on both the technologies and societies that create and implement them)—for lay publics to take seriously their roles in anticipatory governance. The rationale for public engagement implies that such activity can impact those technically oriented and elite actors who write law and policy. UPoT *must devise strategies that compel lay publics to participate in speculative ethics and anticipatory governance.* Otherwise, the gap between makers and consumers will continue to widen, and the consumers will experience a different kind of determinism: elite actors controlling the choices and policies of emerging technologies.

THE NEED FOR SPECULATIVE ETHICS

Like Brey (2010) and Wittkower et al. (2014), Diane Michelfelder (2011) asks her fellow philosophers of technology to address more macro questions. She focuses on the ethical, legal, social, and political implications/aspects (ELSI/ELSA) of emerging technologies.[8] Although there is certainly a need to address specific technologies as they develop, and her work often deals with nanotechnologies and ICTs (2001) that have the potential to transform the human in significant ways, speculative thinking about emerging technologies must continue *independently of* focused attention dedicated solely to *specific technologies*. Reacting to each new development, piecemeal, disadvantages philosophers of technology: our insights depend on particular cases that have already emerged. Michelfelder champions the "need for a speculative ethics of emerging technologies," while acknowledging that her "intent is not to pit targeted, object-oriented ELSI/ELSA reflection against the breadth of speculative thinking" (Michelfelder 2011, 58). Instead, she suggests three reasons to give speculative ELSI/ELSA reflection its due:

> Firstly, speculative thinking adds value to ICT-ELSI reflection because it can allow for important values to emerge that might otherwise go unheeded. Another benefit is that it can open up avenues that might otherwise go unnoticed

> for reframing and recasting issues. It can also, and perhaps most importantly, permit questions to be raised that might otherwise go unvoiced. (Michelfelder 2011, 58)

Her three reasons should serve as a basis for continued speculative work in the philosophy of technology related to the ethical, social, and legal implications of emerging technologies, though here I only have space to examine the last two.

Michelfelder's argument for "reframing and recasting issues," signals the potential for obtaining fresh perspectives on ethical and normative issues from a variety of intellectual resources. Speculative work in philosophy of technology regarding ELSI/ELSA of emerging technologies can leverage resources found in other areas of STS (Guston 2008, 2014). Future fiction, like E. M. Forster's (1909) "The Machine Stops," exemplifies a style writing that depicts future environments that could be created by the incremental adoption of particular, seemingly benign, technologies. Though not all speculative work need be dystopian, as is often the case in future fiction, imagining how humans might manipulate the world around them, and each other, based on the developments of broad schemes of technological developments is worth exploring. As Ethicist Rebecca Roache observes,

> Reflecting on where our most important values lie, and how we might work to maximise them, is surely an important step towards ensuring that ethical concern, and other valuable resources, are not squandered.[9] And, as philosophers have long known, one of the most effective ways of discovering deeply-held values involves speculating about incredible scenarios. (Roache 2008, 326)

Waiting for technologies to emerge before evaluating their potential benefits and harms—as advocated by Nordmann (2007), Keiper (2007), and Nordmann and Rip (2009)—unnecessarily limits the reach of philosophy of technology. It sets ethicists and philosophers of technology, and our interlocutors, at a disadvantage. We would merely react to what scientists and technologists have already developed.

Speculative ethicists do not waste time worrying about potential impacts regarding technologies that may one day be possible. Roache notes that humans lack predictive skill when determining which technologies may one day be realizable (Roache 2008, 323).[10] In fact, Roache offers a compelling argument against such conservative forecasting and assignment of ethical discourse, claiming that there are

> several reasons why it is either impossible or imprudent to restrict ethical debate in this way. First, it risks squandering scientific effort and making it more difficult to contain unethical scientific projects. Second, highly speculative and

> sometimes improbable future scenarios are sometimes instrumental in motivating important ethical projects. Third, the claim that it is always inappropriate to take action to prepare for such highly speculative and sometimes improbable future scenarios is too strong, and contradicts some important ethical intuitions. Finally, the emphasis that some writers place on addressing currently pressing issues may distract us from what is really important. (Roache 2011, 322)

In other words, scientists and technologists at the leading edge of research have no monopoly on determining what technologies may result from their work. Imagining a wide range of possibilities for what might occur, and the underrepresented voices of those affected and affecting technology use, enables a style of participatory governance that examines emerging technologies and their concomitant values. Clearly, a tension remains: multiple perspectives may not prove any more adept at predicting future outcomes than those of the elite actors. However, focusing on the core values that our current and future technologies exhibit permits reflection on, for instance, the type of surveillance we do and do not want and the ways in which our technologies promote and inhibit such surveillance (Wittes and Chong 2014). We need to imagine methods and a language to compare such nonethical values.

Michelfelder's third point, pertaining to raising questions that otherwise might go unspoken, has particular relevance for the lived experience, or phenomenology, of potential technologies. Technologies do not develop in a vacuum; technologies build off other processes, artifacts, and ideas. Similarly, we do not deploy technologies as independent mechanisms that operate outside the bounds of our environments. Michelfelder, describing the future of ICTs, remarks that these technologies

> will continue to intersect and be interlocked with the future of NT, biotechnology, and cognitive science. It is not only advances in these separate emerging technologies that demand to be taken as objects of ethical consideration, but also the overall technological environment resulting from this intersection within which our lived experience unfolds. The fact that the intersection of ICT with NT can lead to the hyperminiaturization of smart machines that, much like the atmosphere, blend into the environment below the threshold of perception, and so create, as Nordmann dubbed it, "naturalized technology," is an example of what I have in mind (Nordmann 2007) . . . seeing what could be learned from the development of the ethics dealing with our responsibilities to the natural world seems a reasonable step to take. (Michelfelder 2011, 58)

Ethical investigation into potential and emerging technologies should highlight our responsibilities to the natural world, including other humans.[11] Speculative and anticipatory governance work has a clear mandate in such investigations. Philosophers of technology—though far from alone in such endeavors—are well positioned to take up these challenges because our work centers on the analysis of human-technology relations. Focusing only on particular instances and artifacts, like instrumental values of particular ICTs and NTs, limits the philosopher of technology and does not allow her work to approach broad theoretical claims because she must instead attend only to what scientists and technologists currently produce. Some—not all—philosophers of technology must consider the "big picture" or run the risk "of missing critically important goods which not only need to be considered in their own right but also can serve as a horizon against which to gauge the threats or benefits posed by ICT-[research and development] with respect to these more instrumental values as well" (Michelfelder 2011, 61).

THE ROLE OF UN-DISCIPLINED PHILOSOPHERS OF TECHNOLOGY

Like Michelfelder, feminist ethicists and epistemologists (cf. Jaggar 1991) have long invited us to acknowledge groups that have not been permitted to voice their opinions and articulate their situations. Lay publics constitute one type of underrepresented group that demand to be heard if we truly wish for our laws and policies to be representative of the people they serve. As Fuller (2010) notes, interactive web forums and focus groups (online and in-person) should be convened to foster education, motivation, and participation in anticipatory governance projects. One recent example, the call by scientists for a moratorium and public discussion regarding genomic engineering and germline gene modification in humans, once again raises the issue of where and how such public discussions can occur (Baltimore et al. 2015). Un-disciplined philosophers of technology have important parts to play in these projects.[12] *How* philosophers of technology communicate, and with *whom* we engage, therefore, must take on an ever-increasing importance.

NOTES

1. Academic philosophy, with its cadre of practitioners that write in a very particular style for a specialized, limited audience, is not organized in such a way that promotes mainstream consumption of its theories and analysis. The UPoT I propose here speculates on and evaluates trends, developments, and concrete cases as they occur, and this work entails moving from the "ideal situations" of traditional philosophy into the messy, and often unclear, scenarios whose best outcomes are equally murky. The UPoT I describe is necessarily informed by STS work. It can be local and practical or global and theoretical, but I see it as meta-STS in important respects. Philosophy of

technology provides a backbone to STS work by giving STS practitioners the means to offer speculative claims despite the seemingly tentative nature of the findings that result from individual cases. It should be approached as both practice and habit.

2. David Guston directs the Center for Nanotechnology in Society (CNS) at Arizona State University, a federally funded academic group charged with providing insight into anticipatory governance of emerging technologies. He (2008) defines anticipatory governance as "a broad-based capacity extended through society that can act on a variety of inputs to manage emerging knowledge-based technologies while such management is still possible" (p. vi). Steve Fuller (2009) describes anticipatory governance as vital to future STS projects and amounts to a "strategy to facilitate the acceptance of new technosciences by inviting people to voice their hopes and concerns in focus groups, science cafes, and computer-based interactive spaces before the innovations are actually implemented" (p. 209).

3. I follow Hans Achterhuis's (2001) demarcation of philosophy of technology, but this distinction also has roots in Carl Mitcham's *Thinking through Technology* (1994). Mitcham distinguishes engineering philosophy of technology and humanities philosophy of technology. Though useful distinctions, his "Notes toward a Philosophy of Meta-technology" (1995) begins to demarcate philosophy of technology in ways that closely resemble how Achterhuis (2001), Brey (2010), and Verbeek (2011b) distinguish *classical* and *modern* philosophy of technology, now a commonly accepted distinction.

4. Lynn Margulis and Dorian Sagan's (2007) *Dazzle Gradually: Reflections on the Nature of Nature* provides an understanding of symbiosis that includes all life on our planet, our cosmos. Meant for a general audience, their work invites readers into a daunting realm of advanced scientific theories by providing copious examples and metaphors that ground science's often highfalutin ideas and technical explanations. Posthumanist positions permeate their work, particularly the notion of technology and the biota as symbionts, a view I enthusiastically support.

5. Wittkower et al. (2014) make a similar claim under a different label: "public philosophy of technology."

6. In conjunction with *Slate*, ASU's CNS seeks to provide—free to those with internet access—lay publics with access to descriptions and opinions of the human-technology impacts of emerging technologies.

7. In a significant sense, this idea seems to miscast the problem. Lay publics lacking sufficient knowledge to inform policy still have much to contribute to the kinds of values they want their policies to promote, and those should affect the kinds of technologies we encourage and incorporate.

8. "Visions and Ethics in Current Discourse on Human Enhancement" by Arianna Ferrari, Christopher Coenen, and Armin Grunwald in *NanoEthics* 6 (2012): 215–29 provides a fine overview of ongoing debate regarding the reach of speculative ethics for cognitive enhancement pharmaceuticals (CEP) and human enhancement (HE) more generally. Further, the 2013 article "STS Policy Interactions, Technology Assessment and the Governance of Technovisionary Sciences" in *Science, Technology & Innovation Studies* 9(2): 3–20 by Christopher Coenen and Elena Simakova makes direct relations between speculative ethics, technology assessment (TA), and STS.

9. The idea of squandering intellectual resources refers to Nordman (2007) and Nordmann and Rip (2009). These texts argue for less research into, and evaluation of, nanotechnology specifically, though their arguments could be extended more broadly.

10. Exponential advances in computational processing power represent just one example Roache gives "of how ideas once thought to be far-fetched have been developed and turned into products that are now a part of our everyday technological environment" (2008, 322).

11. The creation of ASU's CNS, funded by the NSF, makes clear the importance the US government places on the responsible development—to humans and societies—of particular types of technologies, to wit, nanotechnologies.

12. Philosophy of technology should help lay publics confront and evaluate their own tendencies and leanings on topics like the changing notion of privacy and surveillance (Wittes and Chong 2014), not with the goal of convincing people to think in a certain way about an issue, but to give them the toolset they need to evaluate topics and come to their own conclusions with as rich an understanding of what is at stake, now and for the future, in mind.

Chapter 9

What Control? Life at the Limits of Power Expression

Frank Scalambrino

A number of twentieth-century social epistemologists and philosophers of technology deemed our coming twenty-first-century society, with its increasingly cosmopolitan and globalized characteristics, *The Society of Control.* In this regard, Michel Foucault observed "we are entering the age of infinite examination and of compulsory objectification" (Foucault 1991, 200). Technology's relation to such "infinite examination" and "compulsory objectification" is multifaceted. Because technological mediation allows for increased amounts of control, it is as if increasing amounts of examination and objectification emerge as humans cope with increasing amounts of technological advancement. As Jacques Ellul had already noted in regard to *The Technological Society* (1964), technological mediation is far from being a neutral topic of discussion. In fact, Ellul even suggested that technology has acquired a "sacred" status. For example, in teaching a philosophy of technology course, noticing the room filled with students, computers, and smartphones, students tend to quickly chalk up the critique of technology to nostalgia, cynicism, or curmudgeonliness. Invoking fallacious thinking like the appeal to popularity, students tend initially to think that the real question(s) concerning technology merely revolves around whether society would be better without it or not, emphasizing the idea of a mass-scale voluntary relinquishing of technology as an absurdity. Moreover, heated discussions take place in the classroom and online rooted in the assumption that the critique of technology hyperbolically calls for a return to "back-breaking" field work *without a plow*.

In the effort to be as efficient as possible, no doubt, students inquire regarding the most important distinction in the critique of technology, and the answer, I submit, is the cybernetic versus non-cybernetic distinction. Understanding the trajectory of the critique of technology across the history of philosophy reveals how cybernetics alters the traditional criticisms in a radical way (cf. Scalambrino 2014c). So much so, freedom becomes the central issue around which the contemporary critique of technology develops. Yet, because freedom may be understood in a number of ways, it is insufficient to enunciate freedom and control as a self-explanatory binary opposition. Rather, toward a thorough discussion of freedom, it is important to note, at least, that (a) freedom entails, to some extent, control, (b) freedom entails an epistemological dimension, and lastly that (c) existential freedom entails more than physical constraint and temporal quantification.

There have, of course, been multiple moments in history when humans embraced technology as a power that would "shorten the work week" and increase the leisure time of all who could reap technology's benefits. However, today there are now professional assessment and intervention tools designed to help people who cannot seem to "unplug." Just as today the term "addiction" may be used in regard to one's use of technology, for example, cell phones and the internet, so too it is clear that (a) one needs to be able to control oneself if one is to be free. Further, (b) it is clear that without a proper understanding of oneself and the situations in which one participates, it is possible to be exploited. On the one hand, certainly one may freely choose to participate in a situation which others may deem exploitive. On the other hand, it does not necessarily follow that having knowledge of a situation as exploitive means one will not participate in the situation. Yet, knowledge of a situation, as Aristotle's *Nicomachean Ethics* reminds (2009, 1111b4–13a; cf. Scalambrino 2013a), should be considered as a prerequisite for considering an agent to be free in a situation. If for no other reason, consider how a sufficient degree of *self-awareness* is required to know what to arrest for the sake of self-control. Moreover, it is not clear that one can make a choice toward greater freedom if they are not aware of what choices can be made, that is, if the choices required for greater freedom do not appear on the horizon, as possible, given a person's knowledge-base, then this in itself may be characterized as a manner in which the agent is controlled.

Lastly, the "existential" perspective may be the most complicated regarding freedom. On the one hand, one may be said to be free insofar as one is not involuntarily physically confined or insofar as one may spend the time of life, or at least a sufficient portion of it, as one chooses. On the other hand, existential freedom has a greater depth to it. Consider how, given our existence in this historical period, we are not free to exist in other historical periods. What this means, for example, is that we are not free to exist *as* participants in a different historical period; we are not here concerned with attempts to fuse horizons, to act as if, or to pretend to time travel, etc. Yet, this same dynamic may be seen in regard to the historical situation, culture, and society in which we do live. In other words, through something of a coupling of (a) and (b) above toward explicating (c), one may be said to be under control to the extent to which one is not free to be who one is, that is, to take hold of one's own existence in one's own way (cf. Heidegger 1962). In such instances of unfreedom, it would be as if one's existence were carried along by, that is, one were under the control of, a power which had infiltrated to the point of selecting the possibilities available for existential choice without such selection registering in one's awareness.

In the next section, an examination of the cybernetic versus non-cybernetic distinction follows a brief discussion of some critical moments in the history of the philosophy of technology. For it is through cybernetics that one's self-awareness may be infiltrated to the point of unfreedom. Further, despite the trajectory of the history of the philosophy's critique of technology, we find ourselves in the midst of a unique play between objectification and cybernetics. What is unique is that technological mediation seems to have tipped the balance, so to speak, such that the previously mentioned unfreedom is now counted as a kind of freedom. One way to characterize the dynamics of such a play: it is as if the existential unfreedom (c) has shifted into the kind of control taken as a requirement for freedom (a). Such is the context for the following discussion in this chapter. Ultimately, this chapter argues that despite a status as accepted social knowledge, the mere feeling of power associated with progress toward a goal is not to be equated with freedom. Through a history of technological mediation and in a domain of objectification and cybernetics, it is as if the retreat to the feeling of exertion has come to be wrongly counted as evidence of existential freedom.

SOCIAL EFFECTS OF TECHNOLOGICAL MEDIATION: CRITICAL PHILOSOPHICAL MOMENTS

Debates regarding how much the philosophies of Plato and Aristotle truly disagree notwithstanding, we may locate a significant difference, that is, a critical moment, in the history of philosophical attitudes regarding technology between these philosophers. Whereas Plato may be characterized as bemoaning the consequences of technological mediation, with Aristotle we find an expression of the attitude that considers technological advancement and its increasing mediation of (social) reality as necessary. The next critical moment to be introduced is that between Martin Heidegger and Ernst Jünger, and there is something of an analogous relation at work across the critical moments, respectively, that is, Heidegger with Plato and Jünger with Aristotle. Yet by the time of the twentieth-century thinkers, something has radically changed. That radical change is the awakening of cybernetics.

The form of Plato's discussion of the (social) effects of writing should be well known to us. Near the conclusion (274c) of Plato's *Phaedrus*, he stages a discussion between the Egyptian god of *writing*, *measuring*, and *calculation* "Thoth" and "Thamus," the "King" of the Egyptian gods, (cf. Hermes and Zeus, respectively). During a discussion of various means of technological mediation, Thoth describes writing in the following way: "Here is something that, once learned, will make the Egyptians wiser and will improve their memory; I have discovered a potion for memory and wisdom" (274e5). To which Thamus responds,

> Your affection for [writing] has made you describe its effects as the opposite of what they really are. In fact, it will introduce forgetfulness into the soul of those who learn it: they will not practice using their memory because they will put their *trust* [emphasis added] in writing. (275a1–3; cf. Scalambrino 2014a)

The idea here is that writing as a kind of technological mediation, beyond whatever benefits it may confer on its user, simultaneously alters its user. This idea coincides directly with the theme of the relation between technological mediation and changing conceptions of humans. What should be stressed here is that the alterations that follow from technological mediation are not to be simply conceived as a cost or tradeoff of using technology. Rather, it is as if the effects of the technological mediation permeate both the level on which the mediation occurs and a deeper human level. This deeper human level may be characterized as the level of "existential freedom."

A different attitude regarding technological mediation may be attributed to Aristotle. Though Book VI of Aristotle's *Nicomachean Ethics* is often cited regarding his discussion of *technē*, the following position is stated directly in his *Politics*, Book VII. For Aristotle,

> those who have their cities surrounded by walls may either take advantage of them or not, but cities which are unwalled have no choice. . . . [Therefore,] the defenders should make use of any means of defense which have been already discovered, and should devise and invent others, for when men are well prepared no enemy even thinks of attacking them. (1331a7–19)

The idea here is that technological advancement is necessary in that if we do not develop new technologies our enemies will. This is a position often associated with the philosophy of war and militaristic technology and debates regarding violence, sometimes called "preemptive," sometimes called "precipitive" (cf. Wetmore 2005). This attitude toward technological progress and advance belongs to a longstanding philosophical tradition. On the one hand, those who favor this attitude seem to stand on the above argument from Aristotle with merely updated examples. On the other hand, those who resist this attitude emphasize the fact that technological advances, such as those for military purposes, may bring new instruments of greater destruction into the world where previously they did not exist.

Now, of all that may be said of Martin Heidegger's famous statement "Only a God Can Save Us," from the (1966) *Der Spiegel's* interview, it is clear that the statement was made in regard to technological mediation. There, Heidegger notes,

> everything is functioning. That is precisely what is awesome, that everything functions, that the functioning propels everything more and more toward further functioning, and that technicity [*die Technik*, the essence of which is *Ge-Stell*] increasingly dislodges man and uproots him from the earth. I don't know if you were shocked, but I was shocked when a short time ago I saw the pictures of the earth taken from the moon. We do not need atomic bombs at all [to uproot us]—the uprooting of man is already here. *All our relationships have become merely technical ones* [emphasis added]. (Heidegger 1976, 56)

It is in this context, then, that Heidegger's statement makes sense, that is, "Only a God Can Save Us," now that we have become so uprooted through technological mediation. On the one hand, this may be seen as a version of Plato's argument from the *Phaedrus* dialogue insofar as uprootedness refers to the existential consequences of technological mediation, like memory loss. On the other hand, why the ominous tone?

Heidegger explains that something has changed, that is, a threshold has been crossed, which it seems humans will not be able to reign in again. In the sentence following the above quote, he even goes so far to say something which, were it not for the background of Friedrich Nietzsche's "Last Man," it might sound bizarre. "It is no longer upon an earth that man lives today" (56). Heidegger goes further to suggest philosophy itself has been "dissolved" into fragmented disciplines. Hence, we should note, social epistemology may be seen as engaged in the work of recovery through its interdisciplinary discourses. Yet, notice what Heidegger blames for the change, that is, what marks the threshold that has been crossed; Spiegel asks "And what now takes the place of philosophy?" and Heidegger answers "Cybernetics" (Heidegger 1976, 57). This is no longer merely Plato's critique, it is an awareness of the opening onto transhumanism and, ultimately, posthumanism.

Concluding the analogy, then, consider Jünger's discussion of globalization and technological advancement as it presents an attitude different from Heidegger's. Heidegger's *Der Spiegel* interview took place in 1966, and *Technology and the* 'Gestalt' *of the Worker*, that is, sections 44–57 of Jünger's *Der Arbeiter*, was published in 1932. The relation of the dates in regard to World War II is perhaps not entirely unimportant. According to Jünger, "Technology is mastery of the language that is valid in the realm of work" (Jünger 1963/1983, 269). However, "It turns out that [the citizen] is incapable of applying *technology as a means of power* [emphasis added] pertaining to his own existence" (Jünger 1963/1983, 272). He notes that a new epoch is developing which will "devour" the "disintegrating one," and refers to this movement as a "War" and a "decisive attack on individual existence . . . surpassing in significance . . . the Protestant Reformation" (Jünger 1963/1983, 270). Moreover, in the "development toward specific conditions" not only may awareness be publicized of the "imperial unity" of technological mediation, but, further, technological mediation's "final task consists in realizing domination in any place, at any time, to any extent" (Jünger 1963/1983, 276–77). Yet, despite all this, and here is the link to Aristotle, Jünger concludes, "There is no way out. . . . It is necessary to increase the impact and the speed of the processes in which we are involved" (Jünger 1963/1983, 289).

Insofar as these are critical philosophical moments regarding technological mediation, we will revisit them again in the final chapter of this book regarding their ontological import, that is, the changing conceptions of humans and humanity or trans- and posthumanism. However, here, after indicating what is meant by "cybernetics," what remains to be shown is the social effects of technological mediation regarding the organization of social knowledge and its manifestation at the level of individual citizens. It is not entirely unimportant, then, that "Historically, cybernetics originated in a synthesis of control theory and statistical information theory in the aftermath of the Second World War, its primary objective being to understand intelligent behavior in both animals and machines" (Johnston 2008, 25–26; cf. Dechert 1966; cf. Jonas 1953). Etymologically, the term "cybernetics" (from the Greek for "steersman") was introduced in 1947. In his seminal *Cybernetics; or, Control and Communication in the Animal and Machine* (1948), Norbert Wiener explained, "The newer study of automata, whether in the metal or in the flesh, is a branch of communication engineering," and this involves a "quantity of information and [a] coding technique" (Wiener 1948, 42). As such, "Essentially, cybernetics proposed not only *a new conceptualization of the machine* in terms of information theory and dynamical systems theory but also an understanding of 'life,' *or living organisms, as a more complex instance of this conceptualization rather than as a different order of being or ontology*" (Johnston 2008, 31; emphasis added). Hence, by way of an understanding "enhanced" from the cybernetic point of view, the difference between "the steersman's" guiding mechanisms regarding a machine and a living person becomes simply one of calculation and complexity.

What is more, "the complexity of life is not attributed to some ineffable, mystical force" (Johnston 2008, 31). For example, as W. Ross Ashby discussed in *An Introduction to Cybernetics*, the "unpredictable behavior of an insect that lives in and about a shallow pond, hopping to and fro among water, bank, and pebble, can illustrate a machine in which the state transitions correspond to" a probability-based pattern that can be analyzed statistically (Johnston 2008, 31). Hence "cybernetic" may refer to a technological understanding of the mechanisms' guiding machines and living organisms. Heuristically, then, a distinction may be made between theoretical and applied cybernetics: the former refers to the understanding and knowledge of, and the latter refers to relating to, "people," "animals," "things," or "situations" as encoded dynamic systems, that is, ultimately derived patterns of information. What is radical about the power of the cybernetic relation, that is, applied cybernetics, is that when one engages such a relation it is as if the relation engages one back. That is to say, it is as if looking through the all-seeing eye of cybernetics we become part of what it sees, that is, as if we see the steersman seeing us, and yet this self-awareness is, of course, in terms of theoretical cybernetics. Whereas the remainder of this chapter follows out lines of thought in regard to the world ordered in terms of theoretical cybernetics and the manner in which self-awareness of cybernetic control is occluded, the becoming-machine, as it were, which results from applied cybernetics, is followed out in the last chapter of this volume.

POST-DISCIPLINARY POWER: PERPETUAL TRAINING AND THE SLITHERING MARKET CODE

As Michel Foucault no doubt admitted, Gilles Deleuze's sociopolitical analyses went beyond the scope of Foucault's readings (Foucault 1970, 885). Where Foucault was able to speculate, Deleuze was able to articulate, beyond the analytics of disciplinary power. Foucault, following the lead of Heidegger and Jünger, correctly diagnosed the approaching "infinite examination" and "compulsory objectification" (Foucault 1991, 200). However, it was Deleuze who recognized these aspects as indicating an emerging "society of control" (cf. Deleuze 1992). According to Deleuze,

> The socio-technological study of the mechanisms of control, grasped at their inception, would have to be categorical and to describe what is already in the process of substitution for the disciplinary sites of enclosure, whose crisis is everywhere proclaimed. (Deleuze 1992, 7)

As a guiding question: How are such "mechanisms of control" related to objectification and examination, beyond disciplinary power? Notice that what is meant by "enclosure" in regard to disciplinary power is that there are clear criteria to which one may either submit or be punished. By submitting one becomes subject to the identity of the discipline; however, one retains the (inalienable) right to not identify oneself in terms of the rules of the socio-political disciplinary structure, given the privacy of one's own space, that is, *outside* the *enclosure* of the disciplinary structure. Put another way, personifying for ease of expression, in its disciplinary garb power had not determined itself to permeate its codes beyond the public dimension, that is, domain. Hence, the paradigm shift from discipline to control entails a re-coding of the public and private in such a way that, from the perspective of disciplinary power, privacy has been overcome.

Social epistemology is ripe to address the question of how privacy was overcome, because the overcoming was accomplished through social knowledge. Deleuze was already discussing the kind of surveillance surpassing all enclosures that smartphones have made a reality. With a phone's *global* positioning system (GPS) and through the ability to activate its audio/visual input components, we are virtually never alone and virtually never lost. Further, Deleuze notes that with an advancing service industry in the face of constant technological advancement, the post-disciplinary power has encoded the school in terms of the corporation (cf. Fish 2015), thereby ensuring the need for perpetual training. Just as the marketing tactics of today will shift and slide into tomorrow, Deleuze likens the endless pursuit of these goals to the movement of a snake, for example, "control is short-term and of rapid rates of turnover" (1992, 6). "Man is no longer man enclosed, but man in debt" (1992, 6; cf. Mullins 2009), and just as one carries one's financial identity wherever one goes through technological mediation, so too perpetually *in* the steersman's public *eye*, one moves within a paradigm of control.

It is as if the social knowledge that theoretical cybernetics entails inevitably leads to the substitution of private dwelling with public notions of being as mechanisms of control, for example, notions such as "political correctness" and "net worth." As Deleuze sees it, it is as if for the society of control, "know thyself" means "know thy worth," and, according to the steersman, one's current worth depends on variable interest rates, cf. the notion of "epistemocrats" (Fuller and Collier 2004, 21). Because the corporation has infiltrated, that is, permeates through variable rates and an ever-expanding service industry, all levels of what would be private life, debt regulation now functions in relation to a new form (i.e., Jünger's *Gestalt*) of the worker. That is to say, under the control of always already being within a public self-understanding in terms of an ever-advancing ubiquitous technical (i.e., instrumental) rationality that desires a global consumer base, the expression of power itself has become identified as freedom. If theoretical cybernetics offers an exclusive and exhaustive framework for self-identification, then the only remaining freedom would be the freedom to express oneself through the framework or not, though of course, the framework identifies nonexpression as well.

MISUNDERSTANDING NOTHING: IDEAL CONTROL

In his celebrated work *One-Dimensional Man*, Herbert Marcuse was also concerned with control, calling for "freedom from the oppressive and ideological power of given facts" (Marcuse 1991, 227). Noting, "To the degree to which consciousness is determined by the exigencies and interests of the established society, it is 'unfree'" (Marcuse 1991, 227). This is because "Technological rationality . . . impoverishes *all* aspects of contemporary life. . . . There is a *logic* of domination in technological progress" (Fromm et al. 1964). That is to say, according to Marcuse, "The society which projects and undertakes the technological transformation of nature alters the base of domination by gradually replacing personal dependence . . . with dependence on the 'objective order of things'" (Marcuse 1991, 147). Hence, the problem with such objectification is indicated by the very title of the work, that is, "the concept of 'one-dimensional man' asserts that there are other dimensions of human existence in addition to the present one and that these have been eliminated" (Fromm et al. 1964).

The ideal form of control would be to provide a type of self-awareness in which nothing is misunderstood (cf. Scalambrino 2011). As Marcuse already indicated, so long as an individual's needs are being met, technological rationality, or theoretical cybernetics, would seem to provide a sufficient foundation for self-awareness. Hence, no need for self-identification outside theoretical cybernetics seems to arise when the social effects of technological mediation result in the satisfaction of one's needs. According to Marcuse,

> [1st] Independence of thought, autonomy, and the right to political opposition are being deprived of their basic critical function in a society which seems increasingly capable of satisfying the needs of the individuals through the way in which it is organized. . . . In this respect, it seems to make little difference whether the increasing satisfaction of needs is accomplished by an authoritarian or a non-authoritarian system. [2nd] Under the conditions of a rising standard of living, non-conformity with the system itself appears to be socially useless. (Marcuse 1991, 4)

Whereas Marcuse's first point may be seen in terms of the social effects of a cybernetic understanding, resulting ultimately from technological mediation, his second point refers to a self-understanding permeated to the detriment of existential freedom. Yet, because both points involve the satisfaction of needs, might not such satisfaction be understood through an articulation from Louis Althusser? That is, "The ultimate condition of production is therefore the reproduction of the conditions of production" (Althusser 1971/2014, 127).

Insofar, then, that the coupling of a cybernetic understanding with further technological mediation for the satisfaction of needs neutralizes existential freedom, such a coupling results in a twofold accomplishment. First, it allows for the shift to the feeling of exertion to be counted as evidence for existential freedom, that is, the mere feeling of power associated with progress toward a (prefabricated) goal is to be counted as freedom. Second, this neutralizing of existential freedom ensures that the conditions of production are reproduced continuously. Finally, such an understanding situates the normative dimensions of the cybernetic worldview within the scope of social epistemology. For society, and the individuals who people it, a cybernetic understanding of (social) reality provides a framework controlling the horizons of the existential content of life, wrongly affirming a diminished self-awareness at the limits of power expression.

CONCLUSION

It is clear that true freedom entails some degree of control. Moreover, existential freedom as the origin of human freedom is the very condition for the possibility of being under control. That is, only in jest do we accuse a table of being "out of control." This chapter has characterized the substitution of freedom subjected to control for existential freedom as a central feature of what social epistemologists and critical philosophy of technology theorists have called "the society of control." Further, this chapter contributes to illustrating cybernetics as the condition for the possibility of such substitution. Lastly, as the brief discussion of critical moments in the history of philosophy indicated, cybernetics appears along a spectrum of technological mediation, as if effecting a gestalt-like paradigm shift, from which perhaps "Only a God Can Save Us."

Whereas once it was considered offensive to objectify a person, today persons seem to strive for objectification in all areas of life, including, as Marcuse warned, in terms of sexuality and gender (cf. Guillebaud 1999, 279). To the critical reader, such attempts at self-objectification sound like so many appeals to titles etched in the sand by the slithering steersman. Though perhaps advocacy for such inauthenticity may be spotted in all of what Heidegger referred to as the cybernetically fragmented disciplines, psychoanalysis is cybernetic. Moreover, Dasein analysis and Jungian psychodynamic psychology paradigmatically do not belong with psychoanalysis (cf. Scalambrino 2015b). More to the point, psychologists who advocate as if to say "Express yourself! (in a socially acceptable, i.e., publically justifiable, way)" ironically perpetuate the very inauthenticity and compulsory objectification that they claim to be combatting. This is because such psychologists are in the service of the steersman, rather than existence. Hence, they wrongly take being carried along by the steersman, that is, freedom subjected to control, as exemplary of existential freedom.

In a response that does not entirely disagree with Jünger, then, publicizing technological mediation, that is, bringing the social effects of technological mediation into the dimension of public self-awareness, though it may be "nothing" to the steersman, may help develop different dimensions of existence beyond our being subject to one-dimensional cybernetic slithering. What would such experiences look like? To begin, self-awareness regarding the process of selection taking place on the horizon of historical existence should develop. We should avoid using the, ultimately cybernetic, term "unconscious." What is important is not biologically or socially mechanizing the process (or understanding it as such), rather what is important is witnessing our freedom despite a world permeated by technological mediation.[1]

NOTE

1. I would like to thank Dr. John Loscerbo for his friendship and our conversations regarding Heidegger and Technology.

Part II

EXPLORING CHANGING CONCEPTIONS OF HUMANS AND HUMANITY

Chapter 10

Heidegger on *The Question Concerning Technology* and *Gelassenheit*

Charles Bambach

It would be hard to overestimate the significance of the question of technology both for Heidegger's thinking itself as well as for the way this thinking has been taken up and understood within the last half-century. The question concerning technology was hardly a special branch of Heidegger's inquiry. Rather, under the name "technology," Heidegger came to think through the very fate of the West as the destiny of a certain kind of presencing, which had its roots in the work of ancient Greek philosophy and came to shape the way Heidegger understood "the history of being" as a narrative about the possible destruction and salvation of Occidental humanity. Nothing less was at stake for Heidegger as he began to formulate his own careful response to the tragedy of German history and the devastation and brutality it had produced. As such, we need to understand that Heidegger's thinking of technology was very much a part of his postwar attempt to think through the political errors of his earlier works and to offer a singular response to the crisis that affected the German people in the wake of the chaos brought on by their defeat in World War II. Hence, although Heidegger self-consciously avoids any discussion of this *political* dimension in his analysis of technology, the very context and tenor of his lectures on this theme reveal a thinker preoccupied with the question of philosophy's role in the new postwar order of Central European politics.

Although Heidegger had provided an earlier analysis of machination (*Machenschaft*) in the 1930s, he had not truly confronted the relation of technology to his underlying notion of being and to his history of being developed in his encounters with Nietzsche and Hölderlin.

In *Being and Time* (1927) Heidegger had offered an analysis of tools in terms of their utility as equipment and had characterized their mode of being as "readiness-to-hand" (*Zuhandenheit*) that we circumspectly encounter in our surrounding world (*Umwelt*). Later, in *Contributions to Philosophy* (1936–1938) and in his essay "The Age of the World Picture" (1938), Heidegger offered an analysis of scientific representation (*Vorstellung*) and the instrumental thinking that defines truth in terms of its status as a picture that can be placed before (*vorgestellt*) us (GA 5, 89/OBT, 67). As Heidegger came to interpret modernity in terms of Nietzsche's will to power, he saw it as the culmination of a Cartesian subjectivity that harnesses the resources of the planet for the unbridled dominion and mastery of human will over nature. In *Contributions*, Heidegger defines machination (*Machenschaft*) as the "interpretation of beings as re-presentable (*Vor-stellbaren*) and re-presented (*Vor-gestellten*)," whereby beings are taken up only as what is represented and only what is represented is granted the status of being (GA 65, 108–9). This epoch of representability/representation Heidegger terms "the abandonment of/by being" (*Seinsverlassenheit*)—which denotes the most extreme form of "the forgetting of being" (*Seinsvergessenheit*) (GA 67, 7–9, 44, 50).

What shapes this interpretation of Western history as the epoch of the abandonment of being is Hölderlin's myth about the departure and default of the gods (GA 4, 47/64). Taking up Hölderlin's poetic lament about the onset of the "world's night" ("The Gods," v. 7), Heidegger thinks its sense of abandonment in terms of Nietzschean nihilism and the devastation of the earth. For him, this process of the abandonment of/by being coincides with the very history of Western humanity from Plato to Nietzsche; indeed, he underscores that it constitutes nothing less than "the fundamental happening of our history" (GA 65, 112/89). What that means for Heidegger is that *Seinsverlassenheit* comes to be identified with the history of being and/as truth (*aletheia*), of its manifesting itself as presence and of its concealing itself in and through its own modes of self-presentation. At the very heart of this historical unfolding of being as that which abandons us to our preoccupation with and, more emphatically, control/mastery of beings is something deeply paradoxical and enigmatic. Namely, that precisely at that moment in history when the human being has succeeded in harnessing the being's energy and force through its technological ingenuity, has made possible miraculous advances in medicine, transportation, digital communication, and the fields of engineering, architecture, media, agriculture, and the biological sciences, it has lost its simple ability to dwell in proximity to what Heidegger calls "the truth of being." This expresses for Heidegger the grave danger of machination, that by spawning ever more scientific progress and technological production, it succeeds in "disguising" (*Verstellung*) this abandonment and, in so doing, obliterates the very possibility of bringing the dominance of machination into question. In concrete terms, this means that although technology effaces the distance between humans and things in both space and time by bringing everything "near," it winds up producing its own form of remoteness since "everything gets lumped into uniform distancelessness" (GA 7, 167/PLT, 166). Machination, then, produces a way of viewing and representing beings that "in the course of time everything becomes near such that at the same time it loses the character of 'nearness'" (GA 66, 115). On a global level, such machination yields a world marked by devastation (*Verwüstung*) where every possibility of authentic questioning is extinguished.

What emerges from this whole analysis of technology's role in Western history is a profoundly *ethical* question about human dwelling in an epoch of ontological homelessness. Contrary to those who might claim that Heidegger's work can be characterized by its opposition to ethical thinking, I wish to maintain that at its heart Heidegger's thinking is *fundamentally* ethical. What matters above all for Heidegger in his critique of technology is this question of human homelessness in the epoch of technicity. The abandonment of beings by being, manifested in our comportment toward entities as what can be instrumentally configured, designed, and deployed to serve human ends, has become so all-encompassing that it has disguised its own ethical danger to and for humanity. This is not a mere historical trend or passing cultural phenomenon; it is the very destiny (*Geschick*) of the West that has its roots in the history of the forgottenness of being begun in Plato's metaphysics. Hence, we humans are confronted by an essential paradox: technological progress and its global dispensation has brought with it both the greatest indication in history of human dominion over the earth and, simultaneously, the greatest loss of our sense of being at home upon the earth. As Heidegger expressed it in his travel journal, *Sojourns*: "Modern technology and with it the scientific industrialization of the world set about through their incessant proliferation to extinguish every possibility of sojourning (*Aufenthalten*)" (GA 16, 244). It is this harvest of the effects of machination that sets up Heidegger's whole approach to technology.

THE ESSENCE OF TECHNOLOGY

In his famous essay of 1953, "The Question Concerning Technology," Heidegger offers in concise form his mature views on the meaning of technology for the modern era, distilled from his Bremen Lectures of 1949 that were the first public presentation of his work since his teaching ban by the Freiburg De-Nazification Committee (1945–1946). Contrary to the reigning views of technology as a neutral instrument falling under the control of humans, Heidegger offers a radical approach to rethinking not only the essence of technology as such, but also—and more significantly for his purposes—our *relation* to technology. The usual way of approaching technology sees it in two related senses: first, it is an instrument, a means to an end that has no essence in itself; second, it is a "human activity," that is, a tool at our disposal with which we designate ends to achieve our own purposes. Yet both these views—the *instrumental* and the *anthropological*—"make us utterly blind to the essence of technology" (QCT, 4/GA 7, 7). What matters for Heidegger is to pursue the question concerning technology not as a calculus of machinery or systems of information designed for speed, utility, and optimal efficiency (EP, 93/GA 7, 78–9). Rather, it is to pursue this as a question about human *ethos*, the question about how we as humans relate to one another and to the world that opens to us as a place of/for dwelling. Here, Heidegger challenges the common perception of technology as a value-free tool at our disposal that we take up and direct according to our will. On the contrary, he claims, technology is that which *takes us up* in its own way without our even being aware of how the human being inevitably comes to stand within the happening of technology.

Heidegger begins his account by stating that "the essence of technology is by no means anything technological" (QCT, 4/GA 7, 7). Hence, if we cannot free ourselves from the usual definitions of technology, we will "everywhere remain unfree and chained to technology." What Heidegger seeks is to "prepare a *free* relationship to it" so that we may begin to find a pathway out of the instrumental-anthropological definitions provided by the history of metaphysics. In order to do so, he turns to the original interpretation of *technē* in Aristotle's *Nicomachean Ethics* that defines *technē* not as technical/artisanal skill, but as a form of knowledge whose very mode of being is to bring forth truth. Aristotle calls this in Greek *aletheuein*, which Heidegger will think as unconcealment. Yet everywhere Heidegger reminds us that *aletheuein* as unconcealment involves a play or struggle between hiddenness/unhiddenness and is not an assertion "about" entities in terms of a correctness that we perceive. Representational thinking fails here since it thinks of beings in terms of objects for a subject. In Aristotle's account of *technē* as a form of *aletheuein*, however, we become attuned to the phenomenological character of truth as a way or mode of disclosing the truth of beings themselves rather than our assertions about them.

Technē is above all, then, a mode of revealing that lets beings become manifest. Traditional interpretations of technology that privilege instrumentality and view it as a neutral device that serves human ends rely on an understanding of causality bequeathed to them by the medieval church philosophers' reading of Aristotle. There, as the standard teaching tells us, we encounter four "causes" for the production of a silver chalice: the matter (silver) from which the chalice is made (*material*); the form (serving cup) that it assumes (*formal*); the end (sacrifice) which it serves (*final*); and the artisan who plans and serves as the agent who integrates the other three (*efficient*). Yet Heidegger sets out to rethink the four causes as four modes of indebtedness to the process of making whereby these four modes—"all belonging at once to each other"—are understood as ways of "being responsible for" (*Verschulden*) the coming into appearance of the chalice (QCT, 7–9/GA 7, 10–11). Immediately Heidegger clarifies that his use of the terms "being responsible and being indebted" is to be viewed neither "moralistically" nor "in terms of effecting." Rather, they are to be understood as modes of "letting come forth into presencing." For him, the silversmith is not the efficient cause of the chalice, but a coparticipant in a process of unfolding that *lets* the four modalities of revealing bring forth the chalice into appearance. Accordingly, the silversmith acts as the "point of departure" for the bringing-forth (*Hervorbringen*) of the chalice.

Heidegger goes on to characterize these four ways of being responsible as "modes of occasioning" that are "at play within [this] bringing forth." These modes of occasioning he thinks in German as ways of *Ver-an-lassen*: "letting-or-inducing to go forward." *Veranlassen*, then, "*lets* something start on its way into arrival, a being responsible [as] an occasioning or inducing to go forward." To provide some more meaningful context to his reading he cites a crucial passage from Plato's *Symposium* (205b): "Every occasion (*aitai*) for whatever passes beyond what is not present and goes forward into presencing is *poiesis*, bringing-forth." He then goes on to make one of the most important claims in his late work—namely, that *poiesis* is not only a kind of "handicraft manufacture, not only artistic and poetical bringing into appearance." Rather, he stresses, *physis* as well, "the arising of something from out of itself, is a bringing-forth, a *poiesis*. Physis is indeed *poiesis* in the highest sense" (QCT, 10/GA 7, 12). Like *physis*, both ancient and modern *techne* are forms of revealing; that is, both function as ways of disclosing that which comes to presence. Both ancient and modern *Technē* come to presence in that realm where truth, *aletheia*, happens as unconcealment. But Heidegger underscores a profound difference between ancient and modern *technē* that will have serious consequences for the way in which he understands modern technology. Ancient *technē* shows itself as a form of *poiesis*—a way of "bringing forth" that attends to, cares for, and honors *physis*' ways of self-showing. By contrast, "modern technology does not unfold into a bringing-forth in the sense of *poiesis*." The revealing that rules in modern technology is a challenging that alters the way *physis* unfolds and sets it to purposes to expedite, exploit, stockpile, store, transport, distribute and make available for collection and consumption. It places what Heidegger terms an "unreasonable demand that nature supply energy that can be extracted and stored as such" (QCT, 14/GA 7, 15).

Heidegger provides a striking example of different ways of thinking about *physis* in his discussion of the Rhine River. Technology views the river as a source of electric power that can be dammed up and stored for future use, supplying its hydraulic pressure to set the turbines rotating that in turn sets "the electric current for which the rural power station and its electric grid are set up to deliver (*bestellt*) electricity" (QCT, 16/GA 7, 16–17). Here the hydroelectric plant rechannels the river into an interlocking grid of exchangeable power quotients that can be conducted, stored, and converted into consumable energy on demand. In the same way, this technological regulation of the river changes its essential character *as* river and prepares its fate as an object for touristic sightseeing, deliverable upon demand by the vacation industry. What transpires here, Heidegger emphasizes, is nothing less than a transformation of the essence of the river which is now defined in and through the technological grid that imposes upon it "the unreasonable demand" to be nothing other than an "object" there for human control and dominion. He then asks his readers "to gauge the measure (*ermessen*) of the monstrousness that reigns here" and he proposes that we think of the Rhine through a different measure, a poetic measure, that finds expression in the "Rhine" hymn of Hölderlin.

In his lecture course of WS 1934/35 *Hölderlin's Hymns: "Germania" and "The Rhine,"* Heidegger contrasts the scientific-technological representation of the Rhine in terms of "modern science." Such a process "denature(s)" what Heidegger calls "originary nature"; it "dissolve(s) nature into domains of power belonging to the mathematical ordering of world commerce, industrialization, and technology, which in a special sense is machine technology" (HHGR, 178/GA 39, 195). What gets lost in such a process is an originary experience of the river as "homeland," as the historical place of dwelling for a people. And yet Hölderlin's hymn "The Rhine" offers a "poetic thinking" of the "being" of the river that thinks its essence not as energy or supply but as "the mystery of what has purely sprung forth." As a *mystery*, the Rhine remains enigmatic and recalcitrant to human planning and design. But seen as "the principal storehouse for the standing reserve of energy," the Rhine becomes "entrapped as a calculable confluence of forces." Here the *essence* of the river never comes to presence, but is cut off from its originary possibility of self-manifestation, a plight that becomes symptomatic of the calculative representation within modern technology. Instead, the river manifests itself only through the possibilities granted to it by what Heidegger calls "das Gestell": "the technological infrastructure that networks the entire globe of our planet earth."[1]

In Heidegger's idiosyncratic usage, *Gestell* (which in everyday German means a "book rack," "frame," or stand [as for umbrellas]) comes to function as a name for the various ways of positing, placing, composing, positioning, stationing, regulating, adjusting, and setting into the interlocking grid of technological representation. The word itself brings together a range of related cognate terms that all play off the root stem for *stellen* and gathers them (Ge-) within the regimen of positioning. Hence, Heidegger will speak of *bestellen* (ordering), *darstellen* (presenting), *erstellen* (erecting), *herstellen* (producing), *vorstellen* (representing), and *verstellen* (disguising/occluding). As part of the synthetic com-posit-ing of the *Gestell,* all beings get transformed into "resources" made available, stored, and stockpiled as *Bestand* or "standing reserve." Within such a grid there emerges only that which can emerge—namely, beings as disposable entities marked by "enduring availability." As Theodore Kisiel suggests, within the *Gestell* humans are limited to seeing "beings a priori in the horizon of making them useful" (Kisiel 2014, 141). In so constituting all beings in terms of this "challenging setting-upon" that defines the *Gestell,* Heidegger finds the essence of modern technology. What the *Gestell* accomplishes is to unlock, uncover, expose, and extract the secrets of nature in order to make them available for human designation. The revealing that occurs in and through the *Gestell* only allows beings to manifest as standing reserve and hence forecloses any and all alternative modes of revelation. Within the *Gestell* everything is and will be transformed into standing reserve, and with this change comes a fundamental shift in our relation to time. Now for us "time is nothing but speed, instantaneity, and simultaneity, and time as history has vanished from all *Dasein* of all peoples" (IM, 40/GA 40, 41). In this new technological world, we are simultaneously everywhere and nowhere, able to navigate vast distances in seconds and yet unable to abide meaningfully in the very place where we find ourselves. As a result, the human being finds itself at odds with its own sense of place, dis-placed by the placings and positionings of the *Gestell*'s ceaseless placement of human initiative at the dictates and call of ever greater production, efficiency, availability, and distribution. Within the *Gestell* everything is determined in advance by calculations and reckonings that sacrifice all attunement to the ebb and flow of natural rhythms and modulations, for the higher goal of maximal control and subjugation of all that is. Within this overarching, destinal reign of technological supremacy, the human being comes to believe that s/he remains in control of the deployment and consolidation of this technological will to mastery. And yet, Heidegger will claim, this vision of human supremacy over technology remains an illusion, since with the extension of the *Gestell* to all realms of existence, the human being as well becomes a commodity to be stored, shipped, handled, delivered, and disposed of. In the process, as we become the functionaries of technological positioning, we put ourselves in position to be stockpiled and surveyed. As the human being "exalts itself to the posture of lord of the earth," it everywhere experiences only itself—and yet, ironically, Heidegger claims, despite this, "nowhere does the contemporary human being any longer encounter itself in its essence" (QCT, 27/GA 7, 28). Hence, in an age committed to the *Gerede* of diversity, difference, multiplicity, and the radically Other, what reigns instead is the fundamental homogenization of everything that is, according to the compositing uniformity of the *Gestell.*

THE AMBIGUITY OF THE *"GESTELL"*: DANGER AND/AS THE SAVING POWER

What emerges from out of this whole congeries of relations is a world dominated and permeated by the isomorphic imperative to define all beings as commodities available on demand. Concretely, this means that the *Gestell* brings with it merely a single way of revealing beings—namely, as what can be stored, inventoried, collected, shipped, traded, inspected, and consumed. But the greatest threat posed by the *Gestell* is "not simply that it imperils the essence of the human being or its relation to all beings"; rather, as Heidegger puts it, "Where this destining of *Gestell* holds sway, it drives out every other possibility of revealing" (QCT, 27–28/GA 7, 28–29). Moreover, "the dominion of the *Gestell* threatens humanity with the possibility that it could be denied any entry into a more originary way of revealing and thus of experiencing the summons of a more inceptual truth." This, for Heidegger, constitutes "the most extreme danger" of the reign of technology as an epoch in the manifestation of truth. The genuine danger, he reminds his listeners, is not the threat of a third world war or of an atomic explosion that might "bring about the total annihilation of humanity and the destruction of the earth" (DT, 55–56/GA 16, 528). The supreme threat, rather, is that we embrace the calculative thinking that reigns within the *Gestell* as the *only* legitimate form of thinking, and thus foreclose any possibility of what he calls "meditative thinking" (*besinnliches Denken*) (DT, 46/GA 16, 520). This possibility—that we give ourselves over completely to the reign of the *Gestell* "in the sense of the determination of being as the unconditional requisitioning of all beings, including human being"—constitutes the supreme danger since it brings with it the epochal consummation of the forgetting of being that Heidegger had profiled in *Being and Time* (GA 16, 744). Heidegger's great concern here is that with this epochal transformation, human being loses its own roots in being and gets delivered over to the project of technological calculation, thereby losing its own essence and becoming unable to find a proper dwelling upon the earth. Ultimately, the tragic consequence of this whole planetary destiny is that in such a state of rootlessness, with its loss of originary thinking, the human being is rendered incapable of preparing the turn to another beginning for thinking.

Yet Heidegger did not mechanically subscribe to the rants of apocalyptic remonstration. In "The Question Concerning Technology" he pointed to an *ambiguity* within the essence of technology that opened up the possibility of "coming to terms with" it that might occasion a "getting over" or "recovery from" (*Verwindung*) its predominance. This possibility, he maintained, did not lie within our power as lords of the earth, but lay concealed within the essence of technology itself.

Coming to terms with this possibility required that we open ourselves to the way technology manifests its own distinctive way of revealing truth—namely *as* ambiguity. On the one hand, he claims that the *Gestell* "blocks every view into the appropriative event of revealing and so radically endangers the relation to the essence of truth" (QCT, 33/GA 7, 34). On the other hand, the *Gestell* "lets the human being endure (*währen*) . . . so that he may be the one who is needed and used for the safekeeping (*Wahrnis*) of the coming to presence of truth." Ever attuned to the meaning of the appropriative event, Heidegger understands truth as a Heraclitean play of/between concealment/revelation that requires of us that we expose ourselves to a risk: the risk of the unknown, unforeseen, and unbidden power of that which is to come. What this futural coming may foretoken, we humans cannot say; it remains for us a *mystery*. By pointing to this mystery *as* mystery, without attempting to decode its silent force, Heidegger finds the traces of a poetic thinking whose roots go back to Hölderlin and which holds forth the hope of a conversion (*Verwindung*) out of the *Gestell.*

In his lecture of 1949 "The Turn," Heidegger cites the famous passage from Hölderlin's hymn "Patmos":

> But where danger is, grows
> the saving power also. (QCT, 28)

There he lays out a forceful reading of the saving power as that which—far from being an alternative to, or an opposing element of, danger—belongs to danger itself as its ownmost essence.

> Where the danger is as danger, there the saving power also already grows. The saving power does not insert itself alongside the danger. Nor does the saving power simply stand next to the danger. The danger itself is, when it is *as* danger, the saving power. The danger is the saving power insofar as, in its turning, it brings the saving from out of its own concealed essence. (QCT, 42/GA 11, 119)

Heidegger does not turn away from the distress harbored within the being's self-refusal; on the contrary, he patiently shows how the very disguising (*Verstellung*) of the distress brought on by the *Gestell* nonetheless may portend the opening of a possible turn within our historical epoch that remains wakeful for the coming of the other beginning. As ever for Heidegger, the truth of this coming remains a mystery. Hence, he reminds us that we would do well to attend to the meaning of such refusal, since "Refusal (*Verweigerung*) is not nothing; it is the highest mystery of being within the dominion of the *Gestell.*" Since the *Gestell* succeeds virtually everywhere in converting what is mysterious into what can be analyzed, decrypted, disambiguated, and understood, our task would be to preserve the mysterious element within mystery by thinking it *as* a mystery, not reducing it to a quantum of "information." Only in this way can we be open to the futural power of what remains concealed in the first beginning. In Heidegger's words: "As long as we do not thoughtfully experience what is, we can never be owned by what will be" (QCT, 49/GA 11, 77). Ultimately, the saving power unfolds from out of the essence of danger and not through human design or intervention. This is a lesson already prepared within the history of being by the poetic revelation of Hölderlin. And yet human beings occupy an important site within which technology's pervasive force might be held in abeyance and suspended. But again, this involves a stance of ambiguity, of neither completely abandoning technological instruments nor wholly welcoming them. Hence, Heidegger insists, we need to attune ourselves to another kind of thinking than the representational-calculative thinking that dominates technological modernity.

GELASSENHEIT

As Heidegger made clear, there is no simple "overcoming" (Überwindung) of the metaphysical force of technology; rather, we can only hope for a "recovery from" or "conversion" of (*Verwindung*) its domination (GA 7, 77–80). But this conversion—which lies in the essence of technology—cannot be engineered or digitally implemented. In other words, the saving power does not stand outside of technology; rather, this kind of turn can only happen from *within* the essence of technology itself. Nonetheless, technology disguises its essence in every quarter, hiding its mode of revelation in the latest instrumental apparatus available to us that gets looped back into the technical circuitry of the *Gestell*. Yet if we are able to experience the disguising mechanisms of the *Gestell* and to recognize its danger *as* danger, then perhaps we might find ourselves upon a path of thinking that can meditate upon its own entanglement in the labyrinth of technological distortion and obstruction (*Verstellung*). Such a thinking might bring us to the realization that technology is not something to be shunted aside or forsaken, but something toward which—as Heidegger emphasizes—we need to have a "free relation" (QCT, 3/GA 7, 7). Heidegger sees the roots of this free relation as a radical rethinking of our essential stance or comportment to (*Haltung*) technology. This he terms "an equanimity toward things," a way of "letting be" (*Gelassenheit zu den Dingen*) whereby we release our willful ordination of things in order to let them abide.

Within the labyrinth of the technological *Gestell,* Heidegger finds a meaning that conceals itself in mystery. Here, by attempting a stance or comportment of *Gelassenheit*, Heidegger finds a possibility that enables us to keep open the meaning hidden in technology, *openness to mystery.*

> Releasement toward things and openness to the mystery belong together. They grant us the possibility of abiding in the world in a wholly other way. They promise us a new ground and soil (*Boden*) upon which we can stand (*stehen*) and persist (*bestehen*) in the technological world without being endangered by it. (DT, 55/GA 16, 528)

Here Heidegger finds the first indications for a pathway out of technological nihilism in what he calls a "new autochthony (*Bodenständigkeit*)." In this stance of *Gelassenheit* that both lets technological devices into our lives *and* also lets them be outside—that is, "lets them alone"—Heidegger begins to find the traces for the possibility of authentic human dwelling upon the earth.

Heidegger's inquiry into the essence of technology approaches our contemporary condition of "homelessness" in the direction of a more originary site of dwelling for the human being. This he thinks as *ethos*—the sojourn (*Aufenthalt*) of the human being upon the earth, rooted in it and open to the constantly changing, mutable character of human existence in its nearness to being. By attending to this Heraclitean dimension of *physis*, and by thinking it with Hölderlin as the authentic condition for an originary form of poetic dwelling upon the earth, Heidegger comes to understand originary ethics as one attuned to being ever underway amid the shifting haunts and habitudes of human habitation. In radical contrast to contemporary "ethics" which—in the epoch of technicity—defines ethics as a set of rules that dispense normative principles of conduct binding in advance of all human action, Heidegger proposes an originary ethics that speaks to our condition of homelessness. *Ethos* here involves us in the habit of constantly reinterrogating the site in which we are situated. It helps to attune us to the "way"-character of thinking that helps us to let our calculative habits recede in influence and allows us the free relation to technology that Heidegger prizes as the opening of the saving power. Only by entering into this free relation can humans come to dwell poetically upon the earth and prepare the pathway for another beginning for thinking.

NOTE

1. Theodore Kisiel (2014, 134) translates *Gestell* as "syn-thetic com-posit(ion)ing" rather than "enframing" or "positionality."

Chapter 11

How Learning to Read and Write Shapes Humanity

A Technosomatic Perspective on Digitization

Joris Vlieghe

In view of the growing presence of digital media in all spheres of life, I am—as a philosopher of education—particularly concerned with how this affects the way in which we should raise the next generation. I regard education as an intergenerational interaction, and therefore I won't discuss issues like whether writing with pencil and paper is more or less effective than using a keyboard and screen. Practical questions of this kind are without much concern to me, because it takes for granted that in spite of the digital (r)evolution we remain the same kind of subjects, and therefore the only relevant question is whether the replacement of one set of technologies by another has positive or negative effects on education. In this chapter, however, I defend the view that the digitization of education has more severe implications. It might alter the whole framework within which we are used to think about education, and moreover it may reshape our most fundamental conceptions regarding what it means to become an educated person. Taking seriously Kant's idea (1982) that it is *only* through education that we become the sort of beings we are, it is likely that with the digitization of education *a new form of human subjectivity* might emerge.

To make a further introductory clarification, I will not be considering changing modes of attention (hyper versus deep attention) that the use of new technologies might entail (Cf. Hayles 2007) or the possible threat that the absence of real-life encounters with others and the world puts to genuine learning experiences (Dreyfus 2001/2009). Nor will I join those philosophers who have praised (or criticized) the opportunities (or threats) for community building that social media and e-learning platforms entail. Rather, I fully concentrate on transformations that take place at the level of *reading and writing*. This is because what education and upbringing have all been about in, say, the last two hundred years, had first and foremost to do with an *introduction into literacy*. Although the first connotation that comes to mind here is the (basic) ability of reading and writing, I take this term in a broader, more experiential sense: I believe that learning to read texts as displayed in books and to write with pen and paper go together with a very *specific space of experience*, as well as with the possibility to take a unique relation toward the texts that we read and write. And so, with the introduction, proliferation, and increasing dominance of digital forms of reading and writing, things might change immensely.

I. A TECHNOCENTRIC AND TECHNOSOMATIC PERSPECTIVE

Now, saying this might sound rather traditionalist and even utterly fatalistic, as if I claim that today we witness a profound cultural evolution, which moreover will force us to give up something of incomparable value. I agree with the former but not the latter. I must emphasize that my perspective is not normative. This is because I am solely interested in *mapping divergent modes of experience* that go together with the use of different technologies that we rely on in education. This relates to the *technocentric* orientation of my work, which is inspired by the thought of Bernard Stiegler (2008). For him, the word technology has a broader application than mechanic or electronic devices alone. Instead, it concerns *all* tools and supports that we rely on to lead a human life. An electric drill is as much a technology as a prehistoric stone axe is, and the same goes for pen and paper *and* word-processing programs.

In the definition I just gave, the use of the words "tool" and "support" might suggest exactly the opposite of what I have in mind. First, technology doesn't merely relate to things, but also, and more importantly, to embodied practices and gestures related to the use of tools and supports. As such, *typing* with a keyboard and *writing* with a pen are technologies, insofar as they are dependent upon acquiring specific sets of *bodily* routines and disciplines. That is the reason why I call the perspective I adhere to *technosomatic* (cf. Richardson 2010). Second, reading text from a book or a screen is not simply an instrumental activity. Rather, technologies play a *constitutive* role: they are not only things (and practices) that are at our disposal, but also things (and practices) that shape our subjectivity and decide on what we can and cannot do, say, and think. Therefore, *what we are* as human beings is fundamentally dependent upon becoming familiar with particular technologies throughout the years of our education—which could be literally called "formation" (cf. "*Bildung*").

The technologies that "form" us are contingent inventions, and they come and go. Major technological evolutions go together with fundamental shifts in what it means to become educated, and therefore to become a particular sort of human being. However, this possibility of change also applies at a deeper level. Because it is *only* through education that we become subjects, and because the introduction to the prevailing technologies has such a pervasive impact, the existing generation has an immense responsibility toward the "formation" of the coming generations. We have no choice but "taking care of youth and the generations" (Stiegler 2008). Hence, the technocentric and technosomatic perspective I defend in this chapter is far from being *technodeterminist.* Rather, I would say that a correct understanding of the constitutive role of technologies poses a *profound educational challenge* to elder generations, especially in digital times.

II. THE INVENTION OF SCRIPT TECHNOLOGY

In order to develop a more substantial technosomatic account, I turn to the so-called *literacy hypothesis* (defended by McLuhan [1962], Goody and Watt [1968], and Ong [1982] among others). This view partly overlaps with my Stieglerian perspective, but in the end it falls short of giving an accurate account of how technologies of reading and writing "form" us (i.e., constitute the kind of subjects we happen to be). Briefly put, this view sets forth that the technology of reading and writing is not something an ingenious and rationally structured mind invented in order to expand on the abilities that it already possessed in the preliteral, that is, oral culture. Rather, it is the invention of reading and writing that first turned us into the kind of human beings we are today. Our ability for logically stringent thought is dependent upon the possibility to express oral utterances in written accounts (cf. Olson 1977). More precisely it is the linear and consecutive order of written sentences that underpins (cf. Derrida 1978) the clarity and order of what we regard today as a sound argument.

Furthermore, the habit of trusting our ideas to written documents demands the writer to place herself in the position of a nonpresent reader, who has access neither to the context in which the author uttered her words, nor to intonations and other verbal and nonverbal clues (as is the case in oral speech). Therefore the desire came about to have a language that is maximally precise, transparent, and objective, that is, to produce texts that—ideally—speak for themselves. Alluding to the Sapir-Whorf hypothesis, David Olson claims in this connection (Olson 2006, 167) that it is *not so much language that determines reality, as it is script that determines language*. As a result of relying on technologies of reading and writing, our language use transmogrified drastically. We came to realize possibilities that remained hidden as long as language was only used in an oral way. Language became a "schooled language" (Olson 1977, 4): when we are introduced to literacy at school, we "are learning both to read *and to treat language as text*" (ibid., 24, my emphasis).

This is to say, once a text is written down, it starts having *its own existence*. So, when an author revises a text, she isn't merely clarifying what she meant the text to say. Rather, she is confronted with "a visible object, on paper, from which she may infer a meaning that may or may not correspond to her original intention. What remains invariant is not the intention but the linguistic form preserved in writing" (ibid., 24; cf. Barthes 1978). Language thus possesses a force of its own. And we only develop a sense for this potential thanks to some particular education. To illustrate this further, Olson and Oatley refer to what Carl-Gustav Jung, to his own surprise, discovered when he developed word-association tests: "differences between the sane and the insane were minor compared to those between the educated and the uneducated. Whereas more educated subjects gave a single word in response to a stimulus word, the uneducated told stories, offered explanations, provided paraphrases, and the like" (Olson and Oatley 2014, 17; cf. Jung 1910). In order to perform this test adequately, the patient needs to be able to relate to words *as words*—and this is not an ability that is spontaneously given with being a competent language speaker. Rather, it demands a schooled relationship toward language.

Hence, the literacy hypothesis does not claim that literate cultures are superior to illiterate ones or that people who are less exposed to reading are less rational. What this hypothesis *does* imply is that literacy *training* grants the possibility of taking *a unique relation toward the words we produce*, that is, "to treat language as a text." So, the two most fundamental lessons to take from the literacy hypothesis approach are, first, that literacy is more than just the ability to read and write, as it is related to a particular space of experience in which the educated person relates to text *as text*; and, second, that literacy thus defined only results from a specific, viz. *scholastic*, form of education. Here, script is being taught as a goal in itself, that is, in school language is taken "off-line" (Olson and Oatley 2014, 8). Pupils practice reading and writing *for the sake of reading and writing*. Consider that elementary literacy training practices consist of children jotting down, over and over again, the same letter, just in order to get it right. In a sense, many higher forms of literacy education have this typically scholastic character: students concentrate on the *purely formal* dimensions of script, rather than on the practical application of script for "real life," that is, life outside of school.

III. A BRIEF HISTORY OF LITERACY TRAINING

To the so-called *New Literacy Studies* (cf. Street 2003), traditional literacy training practices (and schooling more generally) are looked down upon:

a literacy training that focuses on script for its own sake is, according to this school of thought, as meaningless as studying in detail the manual of a complex machine without ever using it (Gee 2007). Instead, according to New Literacy scholars, we should only focus on *what literacy is good for*. Therefore, it has become popular, for instance, to defend the idea that the only way to secure a bright future for all and to prevent global catastrophes is to rely on "digital literacies" developed outside of school, and more precisely in online "affinity spaces," that is, digital collectives where amateur researchers are gathered on the basis of a shared interest (ibid.). According to the literacy hypothesis, however, this *preoccupation with use over form* all too hastily neglects that in order for a script to have beneficial uses, one first needs to understand the function of written language, that is, that it is possible in the first place to "lay down" (Olson 2006, 176) what we think in material inscriptions, according to a set of conventions (cf. Selcer 2010).

Although I agree with this emphasis on the importance of *formalistic* dimensions of literacy training, I believe that there are serious shortcomings to the literacy hypothesis approach. That is, it fails to explain in detail how a focus on the formal actually *forms* us as a particular kind of subject. In other words, it fails to account for what I have called the *technosomatic* dimensions of learning how to read and write. Furthermore, it could be claimed that the literacy hypothesis approach is, in the end, only about learning to read and not really about learning how to write—whereas the connection between reading *and* writing is crucial. In order to explain this, I go deeper into the concrete, material, and bodily practices of literacy initiation in school, and into the concrete history of educational technologies that lay at the basis of these practices.

Looking at the past seven centuries, it becomes clear that what I have been calling "traditional" literacy training actually dates back to a rather recent date. Many things we take for granted, such as the idea that literacy is *both* the ability to read *and* write, and that a literate person can read/write every kind of text, is solely the result of contingent technological evolutions that took place in the nineteenth century. In late Medieval times, the happy few who got an education were only taught to read. The ability to write was an even greater privilege. Moreover, literacy training was only aimed at "reading" particular texts, such as prayers and songs. In fact, literacy training consisted mainly of starting from a text one knew by heart (e.g., the *Pater Noster*), and then looking at a written version, analyzing its components (Pa/ter Nos/ter, etc.). This assisted students in reciting specific types of texts. However, most of us wouldn't call this activity reading (Chartier 2008).

Literacy only became a general disposition thanks to the invention of syllabaries. These are specially designed handbooks that consisted of rows and columns containing all possible syllables used in a language (e.g., a column composed of ab-ac-ad-af and a row consisting in ab-eb-ib-ob). This meant an enormous step forward, because this particular technology allowed students to become able to read *any* text. Moreover, this also implied that the activity of *learning how to read* became something entirely different and separate from the activity of *reading* itself: whereas the old form of literacy training implied exactly the same kind of "reading" that one was supposed to master as an adult person, learning to read syllables consisted in *training meaningless strings of letters*. Only later on, this skill would be applied while reading meaningful texts. It concerned thus a very specific, and purely *formal* practice that is only to be encountered *in schools* and not in daily (adult) life (cf. Mollenhauer 2013).

Nonetheless, just like in the late Medieval days, learning to read and learning to write remained two separate pursuits. The practice of *simultaneously* learning to read *and* write only arose in the nineteenth century (Chartier 2008, 22). This resulted, again, from specific technological evolutions, such as the invention and spread of resistant metallic quills and cheap cellulose paper, as well as the large-scale distribution of graphic alphabet books. For the first time in history, it became possible to "squander" tons of paper solely with the purpose of *writing* strings of letters like aaaaaaaaa, bbbbbbbb, ccccccccc over and over again. Also, manuals were developed that contained *cursively* printed letters, which is to say that they resembled the rounded and flowing character of longhand writing. And so, due to these developments, the letters one mastered to read became simultaneously the object of rigorous calligraphic practice. Literacy training also became an initiation into writing. With smaller changes, this is still the most popular method of literacy instruction today.

IV. CONNECTION, DISCONNECTION, POTENTIALITY

Whereas the literacy hypothesis claimed that the invention of technologies of reading and writing *as such* brings about a change in the kind of subjects we are, the above brief historical overview has shown that things are more complicated. Much depends on the introduction of *specific* and *historically contingent* technologies of literacy instruction. My point is that in order to understand what literacy, and more generally education, is all about, we need to zoom in on these technologies in their historical specificity, and to take into account material supports and bodily interactions. In this chapter, I argue that "traditional" literacy as we currently know it builds to a large extent on two *unique* characteristics of learning how to read and write that saw the day of light in the nineteenth-century Western school, viz. that it consists simultaneously in *a process of connection and disconnection*, as I will shortly explain. Moreover, I claim that with the adoption of digital technologies these characteristics change drastically, and therefore that we are witnessing today a shift toward a completely different and previously unknown form of literacy.

On the basis of the historical account I just sketched, it could be argued that under traditional conditions, becoming literate is dependent upon a double operation. On the one hand, as correctly stressed by the literacy hypothesis approach, typical to scholastic forms of training is the concentration on formalistic exercises. This is to say that a *disconnection* takes place between the *formal* characteristics of reading and writing technologies, and the *contents* and the *uses* that these technologies facilitate. It seems vital to the acquirement of literacy skills in school that students repeat over and over again strings of meaningless symbols *before* they start applying this skill to deal with meaningful texts. Learning to read and write in this way is qualitatively different from reading and writing themselves. Now, my short historical analysis also shows that this disconnection from real-life purposes goes together with a *connection* at another level, viz. between subject matter (reading texts) and the way in which this subject matter comes into being (writing texts). If one is only able to read without having mastered the skill to produce text oneself, one cannot (or can no longer) be called a literate person. What is at stake in becoming literate seems to be that students simultaneously come and experience, *at a very direct and bodily level*, what it means to generate the very things they learn to read. They literally have "first-hand" knowledge of what *creating* a script is about. Becoming literate is thus not just a matter of consuming letters and words (like being able to drive a car without having any knowledge of its mechanics), but at the same time requires that one has a strong and inside experience of creating letters and words oneself (similar to the experience the car mechanic has while driving) (cf. Stiegler and Rogoff 2010). And it is this repeated focusing of attention on the formalistic, that is, on letters *as letters* (disconnection), which allows reader-writers to stay in touch with their productive powers (connection).

In sum, scholastic literacy training is much more than merely learning how to read and write. What is fundamentally at stake here is that the literate person, thanks to a systematic, prolonged and repeated, and embodied exercising of the formal characteristics of script, acquires an *inside knowledge and familiarity* with what it means to create texts. This productive capability is, so to speak, ingrained in our schooled bodies (cf. Stiegler's [2008] comments on "grammatization"). Drawing from the work of Giorgio Agamben (1999), this closeness to one's own productive capabilities could be called *potentiality* for reading and writing. Potentiality, for Agamben, refers to a strong experience of ability *itself* (a direct experience of oneself as a creature of possibility rather than of necessity). Potentiality is a sense of what one is able to do, which is much more profound than the mere experience that one is able to do *this* and to do *that*. Rather, what is at stake is that one senses that one actually *is able* to do this and that.

Now, what happens when literacy training becomes a fully digital affair, that is, when we no longer acquire literacy abilities by the painstaking practice of repeating senseless strings of letters, but solely by using (touch)screens and keyboards? Looking at this shift from a technosomatic perspective, it could be argued that in training literacy with digital means *a reversed double operation* takes place. On the one hand, there is a *disconnection* between consumption and production of text. Recall that in traditional literacy initiation, the body is being trained in such a way that our fingers form letters that accurately resemble a stereotypical model. Longhand writing is fundamentally a "graphomotoric" enterprise (Mangen and Velay 2010), meaning that writing an "a" and "b" consists of *producing* different things. Over and against this, touching an "a" or a "b" on a screen, or hitting the appropriate keys on a keyboard, is essentially "pointing" gestures (ibid., 396), and as such generating an "a" or a "b" consists of performing the same act. While typing, "the letters are 'readymades' and the task of the writer is to spatially locate the specific letters on the keyboard" (ibid., 386). Even if keyboarding often requires a long and tedious training period in order to acquire necessary bodily routines for typing blindly with ten fingers, the relation between the generative gesture and seeing the letters emerging before one's eyes is completely cut through. No longer is there an intimate knowledge of what it means to produce letters when one points to a key and thereupon sees appearing the corresponding sign on the screen.

At the same time, digital literacy training entails a *reconnection* between form and use/content. Whereas it makes sense during traditional literacy practice to write over and over again the same letter, doing so makes no sense when learning to type on a keyboard or touchscreen. Rather than repeating the same letter, it is vital that one immediately learns to type words and texts invested with meaning: the endeavor at stake is not to learn and type an "a" or a "b," but to acquire an embodied knowledge of which keys to hit in order to make the words one intends to write to appear on the screen. So, for a beginning student, it might be a worthwhile pursuit to type repeatedly one's own name on the keyboard. Here, learning to write is no longer an activity stripped of all meaning: *learning to write is at the same time writing (something)*. A purely formalistic exercise has become an activity with a precise content and use. The same could be said in relation to reading. Learning to read and write with keyboard and screen are no longer an end in itself. Mastering literacy in a digital way is from the very beginning oriented toward meaningful words and texts and no longer deals with the production of letters for the sake of producing letters.

V. DIGITAL LITERACY AS MAKER LITERACY

The conclusion to draw from my technosomatic account is that traditional and digital forms of learning how to read and write go together with *different spaces of experience*. To be clear, I am not saying that digital natives have irretrievably lost something of great educational value. Nor am I claiming that reading and writing on screen would be less embodied. On the contrary, what I argue for is that acquiring literacy in a digital way requires the bodily practice of different gestures and routines, and that therefore *another* and previously unknown sense of being-able, that is, of potentiality (in the Agembenian sense) is involved. The potentiality related to traditional literacy consisted, essentially, in relating to text *as text*, meaning that the literate person has a strong, inside, and bodily engraved sense of what it means to produce letters and words. When the digitally literate, on the contrary, produces letters and words, they are from the very beginning related to particular meanings and uses. The production of letters and words is no longer experienced *from the inside* as creating something out of nothing. Rather, this mode of creation relates to generating new things on the basis of that which already has meaning. This potentiality resides in relating to words and letters *from the outside*. They are experienced as things to "handle."

This experience, of course, is not unrelated to other characteristics of digital technologies, which allow for manipulating and maneuvering already existing texts in ways that traditional reading and writing technologies did not allow for: text is experienced as a computational object, that is, as something susceptible to operations such as cutting, copying, pasting, reduplicating, changing fonts, increasing text size, etc., which instantaneously transform it into something else (instant translation is, inaccurate as it still is, perhaps the most sensational case). Therefore creating is sensed as *building* something new with that which already exists. Using an expression coined by Neil Gershenfield (2012), this sense of creation and of producing things could be termed *maker literacy*. Hence, digital literacy is something substantially different from traditional literacy and should neither be understood in a metaphorical sense (i.e., that it is *like* traditional literacy, be it with the use of different means), nor be gauged in relation to that tradition (i.e., that it is better or worse, more or less effective than literacy as conventionally conceived). Rather, digital or maker literacy is a literacy *in its own right*, as it opens a different dimension of experience, and goes together with completely unforeseeable notions of what it means to create and bring into being "new" things. As such, the digitally educated or "formed" subject is also a *radically new type of subject*.

Chapter 12

Labor and Technology

Kant, Marx, and the Critique of Instrumental Reason

Arthur Kok

Labor is an important factor in human well-being. By participating in the labor community, we are able to satisfy our material needs as well as to acquire welfare and comfort. Since the Enlightenment, the tradition of Western philosophy has become increasingly appreciative of labor as a means to realize freedom, but Marx argues that this quality of labor is undermined by instrumental reason and the mechanization of labor. In this contribution, I explain the core of Marx's critique and I show why instrumental reason and technology can pose a threat to labor as a means to realize freedom. However, I also argue that Marx's presupposition is that labor and reason are means to control nature, but that Kant commits to a concept of labor that can provide a way out of Marx's criticism.

I. ADAM SMITH VERSUS ARISTOTLE

In the discussion about instrumental reason, there are two distinctions that must be taken into account. The first one is Aristotle's distinction between *techne* and *poiesis*. (Aristotle 2009). *Techne* is the kind of rational activity that enables us to create certain desired effects and to avoid undesired ones. It is called instrumental because it is always focused on something outside of itself. It is never the activity that is desirable, but only its result. As such, it is merely a means to an end. *Poiesis*, on the other hand, is a kind of rational activity that is desirable as such, not merely as a means to an end but as an end in itself. The best example of *poiesis* is friendship: the goal of being in the company of others is to be in the company of others. In Aristotle's view, *technē* is inferior to *poiesis* and it can never achieve true happiness on its own. Hence *technē* must be placed in the service of *poiesis*. If *technē* becomes dominant, this will pose a threat to man's happiness.

The second distinction that must be made is the one between *use value* and *value in exchange*. This distinction was introduced by Adam Smith to identify the two ways in which commodities, that is, products of labor, are commonly valued. Broadly, use value identifies the value of a commodity in relation to a specific need. For example, the use value of bread is that it feeds me. Value in exchange, on the other hand, is the value of commodities on the economic market. At the market, it is expressed in terms of *price*. The difference with use value is that value in exchange allows for *comparing* labor products according to general standards instead of subjective ones. Whereas use value essentially is subjective, value in exchange is an intersubjectively shared value (as contemporary mainstream economics argues), or even an objective one (as Smith argued).

Basically, Smith's attempt is to overcome the Aristotelian distinction between *technē* and *poiesis* by showing the positive contribution that technological development—which he connects to economic markets—can make to the realization of the common good. His idea to distinguish between use value and value in exchange has to be situated within a larger historical context in which philosophers tried to conceptualize society using the newly found principles of the natural sciences. Like nature, society is governed by rules that are described in abstract general terms, that is, "natural laws." According to Smith, the natural law that underlies society, defined as the whole of human social interaction, is the *division of labor*.

The division of labor is, say, the general law form of social reality, because it takes the single agent as member of a collective process of production. Labor in general is the act of transforming nature so that needs are satisfied. As such, all labor requires *effort*. This effort can be consumed, that is, it has use value, but we can also trade the fruits of our labor with others, and then it has value in exchange. The value that is being compared is not the ability to satisfy but the effort of making a product, because the effort can be measured as labor time. The objective measurement of products in terms of labor time presupposes a *general* sphere that allows us to abstract from subjective use value in order to be able to compare them with others. This abstraction must be real: we have to be capable of producing more than we actually need to satisfy our needs. Therefore, general value demands a surplus of value that is created in the act of labor. The law that is responsible for this surplus is the principle of the labor division.

The point of the division of labor, for Smith, is that a single agent even in focusing merely on the satisfaction of its private needs is unintentionally participating in and contributing to social reality.[1] As such, it points at a social sphere that is not brought forward by ethical rules or the rule of law, but can be modeled in terms of market economy. Henceforth, the social relations that take place within this sphere are neither ethical nor legal, but *commercial.* This raises the question of how commercial activity relates to ethical and legal action. Smith says very little about this matter other than suggesting that ethical, legal, and commercial activity can exist abreast. However, as Marx points out, the characteristic of individual agents contributing to social life unintentionally creates a tension with the explicitly intentional nature of ethics and the rule of law. How are we certain that ethical and legal action is intentional if unintentional action can create sociality as well? If this is the case, ethical and legal action would not be truly ethical and legal but the continuation of commercial activity disguised as morality and justice.

Essentially, Marx's critique of political economy argues that Smith's theory implicitly makes clear that *technē* cannot be regulated by *poiesis*, because *poiesis* is itself ruled by *technē*. Marx's well-known claim, viz. that the realization of true happiness is impossible under the conditions of market economy, is the result of this fundamental standpoint: the entire idea of *poiesis*, that is, of a domain independent from instrumental reason, is false. Further developing this thesis into an encompassing critique of ideology, Marx argues that also the proclaimed universality of the reason of the Enlightenment is characterized by this false generality of value in exchange. Here, the main characteristic of reason becomes the fact that it is instrumental, which means that Marx's intention is to undermine the reason of the Enlightenment.

II. MARX'S CRITIQUE OF INSTRUMENTAL REASON

Taking the division of labor as the motor of history, Marx argues that reasonable and universal norms of morality and justice are in fact values that correspond to a certain economic system, that is, a process of production according to the principle of labor division, and are reproduced by it. But because they are the result of a division of labor, their universality is only universal to the extent that the principle of labor division is universal. This means, however, that the whole of human action is subsumed under the viewpoint of scientific reason. By appealing to universal principles, science (1) claims to produce objective knowledge, but in fact it is (2) dependent on a specific production process and existing market relations. The universal standpoint of scientific reason implies, with regard to the first point, that the workings of nature are principally transparent to us. With regard to the second one, scientific reason must also assume that it can reproduce the labor that is needed to sustain its universal position by using scientific reason alone. In other words, the claim of objective knowledge presupposes a production process that can reproduce labor entirely with mechanical means only.

According to Marx, it becomes clear that the full mechanical reproduction of labor is impossible in the historical form of capitalism. Marx appeals to the philosophical argument—adopting it from Hegel—that the division of labor is a process of endless differentiation. Every concrete result of the division of labor and every specific mode of production that satisfies a certain demand creates new demands, and this process is one of infinite regression. Scientific reason is characterized by the same regression, which is evidenced by, for example, the fact that every scientific discovery raises new questions. Nonetheless, science adopts the universality of scientific reason under the assumption that we can control nature. Marx argues, however, that the infinite regression to which all scientific knowledge is subjected implies that scientific reason cannot stand on its own feet, but only finds its truth in its implementation in technologies that practically prove that nature can be objectified in natural laws. However, because the production process that is responsible for this implementation is itself *also* subjected to the process of endless differentiation and regression, it does not provide objective grounding. As a result, Marx concludes that the underlying ideal of reason, viz. that we can control nature, is biased as a whole.

The final result of this irresolvable tension is the separation between intellectual and physical labor. On the one hand, intellectual labor represents the position of science that ideologically assumes that it can control nature and see through its workings, as if it were God. The comparison with God is not too far-fetched: the concept of freedom that underlies it is in fact that the production process can be completely reproduced using scientific reasoning only, which means that reason can reproduce itself merely through itself, as an act of creation. On the other hand, physical labor is the actual labor, that is, the hands of the workers, that guarantee that the existing production process is actually reproduced. Their separation entails that from the perspective of intellectual labor, the physical labor is unessential and redundant. It leads to the enslavement of those who depend on physical labor as a means for survival, that is, who are forced into employment for pay even when they lose their freedom by doing so.

Importantly, the prime target of Marx's criticism here is not social-economic inequality in general, but paid employment as a means of capital. Indeed, for Marx, all social-economic inequality is constituted by the introduction of paid employment, but precisely because it reduces labor to something that can be instrumentalized. The introduction of money is already a sign of the dominion of technology. For example, a factory worker who must do assembly line work to be able to survive is disconnected from the fruits of his labor in the sense that the intellectual dimension of the production process is in the hands of a machine. Instead of the laborer employing the machine and determining what it must do, the machine employs the laborer and tells him what to do. In this situation, the laborer is excluded from taking part in the intellectual labor: this part of the production process is taken over by technology. The separation between intellectual and physical labor makes the realization of freedom and happiness impossible.

His critical analysis of the history of mankind as a process of division of labor shows that Marx does not give a critique of social inequality but one of instrumental reason. His critique of ideology precisely makes the argument that a critique of social inequality is meaningless if it assumes a universal principle of justice, because it is exactly Marx's claim that a reference to such universalism could be just as ideological. His argument is that the Enlightenment's reasonable appeal to universality is ideological, because it is depending on a certain state of technological development that is taken to be absolute. Marx's question is how to break out of this illusion of technological supremacy, which means that the separation between intellectual and physical labor must be overcome.

Not denying the possibility of a universal reason altogether, Marx wants to show that capitalism is the penultimate state before freedom and happiness can be realized in the actual world. There is reason in history, Marx argues in line with Hegel. Yet for Marx, this reason remains a hidden force—an invisible hand—until capitalism is overcome. Exactly its necessary hiddenness under the conditions of capitalism means that reason cannot actualize itself *self-consciously* through the actions of the people. Instead, capitalism worships a false deity, viz. the freedom of the market. Capitalism must be radically abrogated by means of a revolution that makes explicit the historicity of reason. Although it remains rather undetermined what the positive result of this revolution, that is, communism, should be, it is clear what Marx's philosophical conclusion eventually is: under market conditions, labor can never lead to freedom, because these conditions have the form of a *natural* and *unintentional* process, and free labor has to be *self-conscious* labor.

III. A KANTIAN PERSPECTIVE ON LABOR

Interestingly, the greatest protagonist of Enlightenment's reason, Immanuel Kant, would agree with Marx that labor, which can realize freedom, must be self-conscious and cannot be subjected to the rationality of the market. However, Kant would not accept that labor is necessarily subjected to instrumental reason and to the division of labor. On the one hand, when Kant speaks about economic labor in the *Metaphysics of Morals*, he does not use the common German word for labor, viz. "*Arbeit*." Instead he uses the word "*Fleiß*," which means diligence or industriousness (Kant 1996, 286). On the other hand, Kant often uses the word "Arbeit" in relation to the activity of philosophy itself. He regards philosophy as hard work, but it seems implausible that he thinks about common labor in this context.

In the *Critique of Pure Reason*, Kant jokes about the relatively small payoff of philosophy in relation to the tremendous effort it requires. Philosophy is the activity of becoming self-conscious (about one's presuppositions), that needs to be undertaken for the purpose of becoming self-conscious and not for economic interests. There is one payoff, however, that can be made "comprehensible and interesting to even the dullest and most reluctant student," viz. that he can avoid with certainty "many embarrassing corrections" that result from the uncritical use of speculative reason (Kant 1998, B 297). So for Kant, there is value in knowing, but only in *absolutely* knowing the boundaries of understanding: the value of not constantly being afraid that someone will point out that you are wrong.

This is interesting in relation to Marx, because he argues that the value of knowledge is the one that is objectified through the division of labor in the production process. Obviously, this is the case for positive scientific knowledge, but Kant stresses the value of knowledge of the *limits* of reason. In fact, he even suggests that this value is a source of *coercion* in the sense that it can compel someone with no intrinsic motivation to do philosophy. How should we conceive of this kind of coercion that is clearly not one of economic necessity? In his *Anthropology from a Pragmatic Point of View*, Kant links this kind of coercion to labor, explicitly disconnecting it from economic necessity. He says here that coercion is not against human nature, but that humans even desire tasks that demand a certain effort to accomplish. He illustrates his point by suggesting that even Adam and Eve in the Garden of Eden would not just "sit around and sing Arcadian songs," because "boredom would have . . . tortured them" (Kant 2007, 471).

The point is that Kant's concept of intellectual labor is different from Marx's because it does not relate to the division of labor at all. The assumption of the division of labor is that labor activity consists in the transforming of nature, giving it the shape of consumable products. Intellectual labor, in this conception, articulates the knowledge that is necessary to perform such transformation. The labor to which Kant refers in his quote is intellectual labor because it stems from an inner motivation as its source of coercion. When we link this to the remarks from the *First Critique*, this inner motivation is explicitly the motivation of a self-conscious being, that is, the human being.

Marx would interpret this Kantian separation between economic and intellectual labor as ideological. In Kant, the main difference with economic labor is that intellectual labor is *free*. Marx would not dispute Kant's claim that free labor must be self-conscious, but he would argue that Kant's position is ideological because it does not do what it claims to do: Kant intends to make possible a free and self-conscious labor activity that contributes to happiness, but in reality he reproduces the repressive forms of the social and economic reality of his society on an abstract level. The freedom of Kant, which he identifies as the noumenal realm, is therefore only abstract. From Marx's perspective, the separation between the phenomenal and noumenal world means that freedom is cut loose from the material world. This not only results in the illusion that freedom can exist without the physical world and its bodily needs, but also excludes the physical world from participating in the intellectual world. Yet the exclusion of materiality is the same as the false belief that nature can be completely under our control and that technology can mechanically reproduce itself; and therefore, the kind of intellectual labor that Kant derives from it is also abstract freedom.

However, to argue that Kant's idea of independence from the material world is not actual freedom but merely an abstraction of a social and economic reality relies on the assumption that Kant's abstract freedom can be identified with the abstract freedom of the market. Hence, it must be shown that Kant's separation between the phenomenal and noumenal world can be reconstructed as an outcome of the abstraction that takes place under the division of labor. But this is impossible. The separation between the noumenal and phenomenal world cannot represent the one between intellectual and physical labor, because the first separation only exists for someone who is self-conscious about the limits of scientific reason. Marx's criticism was exactly that capitalism makes free labor impossible, because labor could not become self-conscious. Kant's two-world theory thus cannot be ideological for the simple reason that it is the reality of self-consciousness: to be self-conscious *means* to be aware of this distinction.

IV. CONCLUSION

The presupposition of Marx's critique of ideology is that the purpose of reason—and henceforth that of science and technology—is control of nature. He makes this explicit in his approach to history that takes the division of labor as the motor of human development. As a result, freedom and happiness depend on an instrumental process for their realization. In capitalism, this immanent logic of history becomes explicit in the absolute separation between intellectual and physical labor, making visible the absolute impossibility to reconcile technological development with human freedom. However, this presupposition cannot be applied to Kant. For Kant, the necessity for labor refers to becoming self-conscious as such; recall his conception of labor activity that involves coercion but explicitly a noneconomic and noninstrumental one—it exclusively does not refer to the division of labor.

Moreover, in a way, Marx might be more Kantian than he realizes. Though he does not accept the irreconcilable difference between the noumenal and the phenomenal world, he believes that the only way to overcome it is through revolution: the radical sublation of the existing order. But then he is not so much criticizing the Kantian distinction between empirical and intelligible world as an ideological representation of social-economic reality, but rather redefines it in terms of social-economic reality: the noumenal world is the reality of communism, in which the disruption of the empirical world—the existing social-economic reality of the market economy—is overcome. So if there is any transition from a more abstract to a more concrete approach in the development from Kant to Marx, then it is the fact that Marx takes up Kant's metaphysical reflections and turns them into a theory for political action.

NOTE

1. This refers to the most famous quote from Adam Smith, *The Wealth of Nations*, "It is not from the benevolence of the butcher, the brewer, or the baker that we expect our dinner, but from their regard to their own interest. We address ourselves, not to their humanity, but to their self-love, and never talk to them of our own necessities, but of their advantages" (2000, 15).

Chapter 13

The Biopolitics of the Female

Constituting Gendered Subjects through Technology

Danielle Guizzo

This chapter discusses the role played by technological artifacts in reinforcing biopolitical forms of control and regulation that have the female body and gender as their main targets.

French philosopher Michel Foucault (1926–1984) opened many possibilities for a rethinking of the condition of subjects within modern society, conceiving them as a product of a specific time and place. Among his valuable contributions to the development and critique of human and social sciences, Foucault developed and improved the notion of biopolitics, a technology of power that acts over life and its biological processes in order to regulate and normalize populations. Almost forty years after Foucault's first writings on this subject, the issue of biopolitics still remains a contemporary topic, particularly for those who are interested in the social construction of subjects and how they constitute themselves according to certain rules of power in contemporary society.

One plausible focus that demonstrates the originality and contemporaneity of biopolitics is the field of gender studies and the construction of the female body and gender. As Foucault demonstrated in *The History of Sexuality Volume 1* (1978), the practice of biopolitics reinforced many forms of control and the administration of a variety of issues that were connected to sexuality, which created a *scientia sexualis*. This new form of knowledge, as Foucault stressed, created medical, discursive, and moral apparatuses that had as one of its main targets the female body and gender.

In the context of creating the female, one must consider the strategic role of technological artifacts in reinforcing biopolitical forms of control and regulation. Here, Science, Technology, and Society (STS) studies may provide a fertile contribution to investigate how the female condition is reinforced by technological devices used by medical, educational, and moral institutions (for instance, see Wajcman 1991; Bijker et al. 1993; Hopkins 2009). Indeed, many emerging technologies can actually contribute to reinforce the condition of the female gender insofar as it creates new normative patterns, such as the issue of reproduction and its control.

The next section discusses the philosophical foundations and political implications of biopolitics and power relations with special emphasis to Michel Foucault's contributions and his contemporaries. The second section delimits the field of STS studies concerning the critical study of technological devices and their implications for society, focusing on the STS approach to gender issues and its political and social implications. The last section summarizes the constitution of the female through technological artifacts and how this actually constitutes a specific biopolitical form of power inasmuch as technological artifacts act in accordance to biological forms of control and regulation.

I. FOUCAULT, BIOPOLITICS, AND THE REGULATION OF POPULATIONS

Michel Foucault's investigations involved different areas of human, social, and medical sciences, such as psychiatry, sexuality, clinic, law, economics, philosophy of science, discourse, and language, among others. His vast oeuvre included books, courses, interviews, and several lectures, which were not separated into isolated phases, rather in theoretical displacements[1] inside the same perspective: to account for how and why individuals became subjects in modern age.

The issue of power—as well as its creation, forms of dissemination, and possible consequences—became the focus of one of Foucault's axes, more specifically a phase called genealogy of power. With the influence of Friedrich Nietzsche's works on the genealogy of morality, Foucault developed a genealogy of the governmental practices that occurred from the sixteenth to the twentieth centuries to understand the existent power relations between subjects and institutions; more specifically, the state and other social institutions such as hospitals, armies, schools, and factories (see Foucault 1978, 1995, 2004, and 2008).

During his investigations, Foucault noted how powers could be individual (certain types of power that have an isolated individual as its target) or collective powers (those powers directed to the population). When dealing with collective forms of power, Foucault acknowledged the role of the state as a spreading force of powers, particularly the modern institutional framework of the state insofar as it created and disseminated new forms of power relations throughout modern age (see Foucault 1978 and 2003).

Foucault claimed that during the eighteenth and nineteenth centuries, a new form of power emerged that was not a punitive one. It was a type of productive power that focused on human life as a whole: biopolitics (2004, 1). For Foucault, biopolitics was a technology of power that emerged at the end of the eighteenth century and aimed at organizing power over life in its collective way. Insofar as it dealt with several biological elements of human beings, such as birth, mode of living, prosperity, health, reproduction, and death, biopolitical practices demanded a set of strategies and techniques in order to assist in controlling and maintaining human lives. These techniques included statistical devices, demographic control, design and implementation of public policies, and discourse legitimacy, all of them becoming more sophisticated and intense throughout modernity.

It was during that time that the subject started being considered not individually, but mainly collectively, emerging as a *population*. The transition from the individual to the population as the embodiment of a political subject occurred throughout the eighteenth century, a period of time regarding which Foucault claimed one of the first great novelties in power techniques appeared, that is, the framing collectives of the human population as an economic, social, and political issue (Foucault 1978, 31). At that historical moment, governments realized that they could not manage isolated individuals, but a population with its own regularity and specific variables: birth, life expectancy, reproduction, productivity, habitat, and death.

Foucault's book published in 1976, *The History of Sexuality Volume 1* (1978), demonstrates how the practice of biopolitics also reinforces many forms of control and administrates a variety of issues connected to sexuality. For example, reproduction is a key issue in controlling and managing populations, hence the emergence of medical, discursive, and moral apparatuses that had the female body and gender as its main target. That is precisely why Foucault stressed the denial of a repressive hypothesis about sexuality and emphasized the normalization, utility, and surveillance of bodies through many public and private policies.

Contemporary readers of biopolitics such as Dean (2010), Lemke (2011), and Miller and Rose (2008) underpin the relevance of biopolitics for understanding current medical, educational, urban, and economic policies, particularly due to their role in constituting conducts and shaping subjects. We cannot think about power without considering the role of social institutions, but technological artifacts also have an important part in effecting biopolitical practices.

In this sense, Foucault (2004, 11–5) draws attention to the role of security apparatuses (or *dispositifs*) as practical instruments and mechanisms of action of biopolitics. A security apparatus can be characterized as a technique, a technological device, a discourse, a calculus, or a policy. It acts over a space constituted by a multiplicity of subjects and aims at normalizing and balancing all types of problems that appeared in the process of governance. Put differently, a security apparatus is the means of action of biopolitics that acts as a practical form in regulating and controlling populations.

For instance, Foucault (1978, 46) exemplifies the case of social medicine and the concern with sexuality in the nineteenth century: when sex became an issue of populations and social medicine, new medical and social measures that involved female and infant bodies emerged, such as the care of babies (breastfeeding, hygiene), the segregation between boys and girls in schools, and the polarity of parents and children's rooms. Foucault noted that this latter segregation actually determined the architectonic model of house building throughout the nineteenth and twentieth centuries.

In that context, one can realize the role of mechanisms, techniques, and also technological artifacts in developing new forms of control that are not necessarily repressive, but productive: they produce knowledge bases, behaviors, and patterns that do not resemble punitive modes of living. The next section focuses on the social construction of technology and its impacts on life, particularly on the constitution of gendered subjects.

II. SOCIAL CONSTRUCTION OF TECHNOLOGY AND ITS IMPACTS ON GENDER STUDIES

STS studies emerged in the decade of 1970[2] as a critical school of thought that questioned the role of technological artifacts in human life and its alleged benefits to the development and improvement of life quality, jobs, and cultural relations.

STS scholars such as Latour and Woolgar (1986), Callon (1993), Pinch and Bijker (1993), and Feenberg (2002b) emphasize the social consequences of technological devices from an interdisciplinary and critical perspective that combines (but is not restricted to) history of technology, philosophy of science, historical materialism, and sociology of scientific knowledge. STS intellectuals reject the statement of scientific neutrality and technological determinism, claiming science, scientists, and technological artifacts can affect human behavior, culture, social, and power relations.

Indeed, as Sismondo (2010, 57) stresses, *social constructivism* provides three important assumptions for STS studies. First, that science and technology are, above all, *social*. Secondly, they are *active*, which implies that science and technology play a relevant role in constructing new patterns. Lastly, science and technology *are not natural* in the sense that they do not provide thoughts and conclusions about natural, but rather socially constructed realities. Together with those forms of socially constructed realities, social groups also determine the social and cultural production of science and technology. Pinch and Bijker (1993, 30) define *relevant social groups* as institutions, organizations, or groups of individuals—either organized or not—and demonstrate how their influence can determine the success or failure of a technology based on its social acceptance.

In that particular context one cannot ignore how technological artifacts are nonneutral in the sense that they affect individual and social life deeply. If technology had a neutral character, social and environmental impacts would be simple accidental side effects of progress. However, following Feenberg's instrumental[3] analysis that underpins the nonneutrality of technology, clearly choosing a certain technology actually has political and social implications inasmuch as technological artifacts carry social values themselves. Therefore, technology does not cause labor or environmental degradations, but antidemocratic values that govern and determine technological development. A possible alternative that can reverse this antidemocratic process is the democratization of technology, thus allowing equal political, economic, and cultural access to the design and implementation of artifacts in order to avoid conflicts of interests.

A more specific approach of STS studies that focuses on the perspective of gender issues is the field of *gender and science* (see Keller 1995b, 82). Oudshoorn (1994 and 1999), Keller (1995a), and Wajcman (1995) demonstrated that science and technology can also affect the construction and shape of the female body and gender in a very intense way, mostly due to the social, moral, cultural, and power influences that act behind technological artifacts. For instance, as Oudshoorn points out, the emergence and constant improvement of birth control methods only became possible due to a construction of a dualistic male/female notion of sex hormones between 1905 and 1920 by sex endocrinologists. Quoting Oudshoorn, "This conceptualization not only harmonized with the prescientific idea of a sexual duality located in the gonads, but also found a ready acceptance given the cultural notions of masculinity and femininity of the day" (1994, 21–22). In this sense, medical theories about the female body actually reinforced the binary medical treatment of males and females, especially concerning sex hormones, which led to the development of technologies that emphasized this gender difference.

Exploring the emergence of birth control pills in the light of gender and science studies provides us a bigger picture of how technologies can affect social and cultural norms. At first, birth controls may be considered as a technological artifact enhancing female's independence and power of her own body and decisions. Furthermore, we cannot ignore the fact that contraceptive methods entered the standard agenda of development process in several underdeveloped countries in the past decades (see van Kammen 2003, 151). Nevertheless, as gender and science scholars claim, this sense of independence was annulled at the moment that some institutions, such as medical, family planning, religion, and the state, began regulating and controlling the use of birth controls, leaving the administration of the female body and behavior to other individuals and/or institutions.

If technology is not neutral and it acts according to a particular hegemony of thought, as Feenberg (2002b, 63) presupposes, then all the decisions made under that paradigm tend to reproduce that hegemony and reinforce the power of interest groups. In this sense, Feenberg (2002b, 68) takes Foucault's notion of power relations to demonstrate how knowledge and truth are a product of power relations and influences that are disseminated within the social reality. In this sense, technology arises as a mechanism or an instrument of social dominance insofar as it helps in the exercise of power through control, normalization, and regulation methods.

Therefore *a proper understanding of the power forces that influence the behavior and administration of female bodies in the contemporary age should not be dissociated from the mediating role of technology in the process of life or the scientific interests and institutions involved in such mediation.* That is why one should think about a biopolitics of the female, a specific technology of power that creates a form of knowledge, a discourse, and truth systems with the purpose of shaping and regulating the female human body, thus hegemonically establishing the gender of the "woman."

III. FEMALE AUTONOMY OR GENDER CONTROL? THE ROLE OF TECHNOLOGICAL MEDIATION IN LIFE

Technology is an instrument of multiple forms of domination that act according to certain interests, such as male domination of women and nature (Wajcman 1995, 194). Indeed, one considering technology as an instrument or a mechanism actually presupposes an ideological, historical, social, and cultural background for its use. The case of technologies that aim at the female body and/or gender are not different from other artifacts insofar as they also constitute identities, behaviors, experiences, and discourses that go beyond the reproductive context. For instance, several technologies that involve female labor and domestic activities such as domestic supplies (cf. Cowen 1983) also reinforce the power relations involving women. Although they may represent a "solution" to women's oppression in the home, we often ignore the fact that they can increase the gender bias involved in domestic activities.

Technological artifacts that are designed with the specific purpose of regulating and controlling the female body and gender act as security apparatuses, or positive forms of improvement and secure individuals in order to make their lives better. As Foucault (2003, 242–23) and Lemke (2011, 36) underpin, biopolitics differs from disciplinary forms of power that aim at controlling and repressing individuals by creating norms, surveillances, exclusions, and disciplines. Biopolitics does not aim at repressing and disciplining individuals. On the contrary, biopolitics has a regulating character that intends to normalize and equilibrate populations by using the principles of security and stabilization, whose use began at the end of the eighteenth century and still remains today.

In that context not only populations in the collective sense of the term emerged as political subjects, but also the woman—both her body and gender—emerged as a political subject. Foucault's analysis of the *scientia sexualis* can be reinterpreted from a contemporary perspective in which medical, social, cultural, and economic institutions are responsible for separating male and female bodies and creating gendered subjects with specific tasks and normative behaviors. Technological artifacts and their biased design act as an instrument or apparatus to help disseminate positive and productive forms of power, that is, biopolitics.

Perhaps one of the most controversial biopolitical forms of technological mediation involving female body improvements (cf. breast implants and vaginal plastic surgery) and binary gender production is the *birth control pill*. Although it deals with the possibility of female empowerment and self-control, the pill also combines strong medicalized, cultural, and social elements that reinforce her body and hormones as a place of social and power intervention, making her the only one responsible for the issue of reproduction. Following the notion of biopolitics, *the birth control pill indeed represents a technology over life that creates and disseminates new forms of power relations influencing how one should act or behave, going beyond its strict biological purpose*. As Oudshoorn stresses, "The pill was a novelty since it would be the first drug in the history of medicine given to healthy people for a social purpose" (1994, 111; cf. Bray 2009).

The discourse provided by scholars of technological determinism implies the uncountable benefits of technological evolution, claiming that major technological advances transformed women's lives for the better, such as technologies of pregnancy and childbirth, as well as reliable forms of contraception and abortion, which meant an autonomy of women's bodies for the first time in history (cf. Wajcman 1991, 54). Nevertheless, the cultural, social, and historical backgrounds, that is, genealogies, of such technologies are often ignored, just as the power relations and structures that surround the issue of the female body and gender. Reproductive technologies such as *in vitro* fertilization and birth control pills represent a biosocial imperative that acts over bodies in the form of a biopolitical control (cf. Whetstine 2015).

For some gender and science scholars (cf. Keller 1995a and 1995b; Wajcman 1991 and 1995), this creation of gendered artifacts can be explained by the male-patriarchal domination of technology and its consequential desire for control and power over the female body. Indeed, the argument should not be ignored when approaching the design of a nonneutral artifact such as the pill. However, one should realize that the male-patriarchal account of the domain of technological mediation is not enough to explain and justify the expansion of several public policies which use the birth control pill as a form of regulation of mass fertility and how it is socially accepted by women and considered a way of female autonomy. Thus, Foucault's notion of biopolitics can demonstrate the use of technologies as an instrument that disseminates certain forms of power relations throughout society with a dual aspect of control and autonomy.

As Foucault (2008, 1) accomplished an investigation of biopolitics and its consequences, he stressed how the effectiveness of biopolitical power depended on the spread of certain knowledges or discourses to create systems of truth. The specific case of technological artifacts designed for a gendered construction of the woman—such as domestic appliances and sexual and/or reproductive medicaments—demonstrate the effectiveness of artifacts as security apparatuses that function according to biopolitical power and go toward rationalizing the "problems" posed for humans. What is astonishing is that these machinations actually constructed a discourse based on female autonomy, independent decision making, and self-control, thereby legitimating the intervention on female bodies with the purpose of regulating their sexual and reproductive functions (cf. Anderson 2014, 19; cf. Jeffreys 2008).

This politicization of life that encompasses technological mediation and the use of technological artifacts specifically designed for the female body also enhances a gendered notion of women. Increasing interventions on sexuality, reproduction, pregnancy, female labor, female behavior, and female aesthetics represent the dissemination of a biopolitical power that uses technological mediation, that is, technological artifacts as mechanisms (or apparatuses), to guide and determine "normal" women into an accepted pattern of living.

IV. CONCLUDING REMARKS

Biopolitics remains, in general and specifically for social epistemology, a relevant concept with which to understand how productive forms of power spread over society and influence our behavior and constitution as subjects. Foucault's analyses of the influence of power relations in the lives of individuals emphasized the role of biopolitics as a technology of power that acted over populations with the purpose of constructing productive, normal, safe, and regulated subjects. In that context, female bodies also became one of the objectives of biopolitics insofar as the female is strictly connected to reproductive, sexual, and domestic issues.

Hence, where biopolitics exercises its domain, technological artifacts emerge as a security apparatus that aids in the dissemination of powers and discourses that affect the biological nature and social behavior of females. Making use of interdisciplinary research programs, such as social epistemology and STS studies, can offer us a critical perspective regarding the nonneutrality characteristic of technology, which perpetuates biased views of several issues related to a gendered woman, such as sexual, reproductive, labor, and aesthetic patterns.

Associating Foucault's investigations of biopolitics with the gender and science approach leads to some interesting insights for a rethinking of the forms of power exercised in our world today and the role of technology in life, which usually presents itself as an "improvement" with a "free" character. For instance, if one looks closer to the rise of the birth control pill as a technological device and the many discourses surrounding it, we may notice that the pill is usually construed in terms of female autonomy and empowerment, that is, self-control over her own body and free decision making. However, in terms of social epistemology and technological mediation, the pill is a mechanism that disseminates a productive form of power affecting the biological functions of a woman, and thus may be seen in the biopolitical service of reinforcing her gender condition.

NOTES

1. Foucault's three theoretical displacements include: (i) archeology of knowledge; (ii) genealogy of power; and (iii) ethics of the subject. This chapter deals with the second one.

2. Although many studies involving the social consequences of technology were published during the 1960s, we consider the recognition and spread of STS as a proper field in 1975 with the foundation of the Society for Social Studies of Science.

3. Feenberg's (1995 and 2002b) approach to technology has an instrumental character, where he (2002, 5–7) treats technology as subservient to values established in other social spheres (such as politics and culture). According to Feenberg, his approach differs from the substantive theory of technology found in Jacques Ellul and Martin Heidegger in the sense that they treat technology as fatalism and as an autonomous cultural force.

Chapter 14

Phenomenology of Radiology

Intentional Analysis in the Constitution of Diagnostic Judgment

Mindaugas Briedis

This chapter brings together two fields of research and practice which, prima facie, seem methodologically unbridgeable, that is, philosophical phenomenology and empirical medical diagnostics through radiological images. Whereas such diagnostic judgments result from technological mediation, this chapter situates these judgments within the framework of what is phenomenologically understood as "intentional analysis." Taking Edmund Husserl's phenomenology as a point of departure, the role that imaging plays in the constitution of the intended object, the role of embodiment, imagination, and empathy is explored. As evidenced by a Husserlian reading of intentional analysis in the constitution of diagnostic judgment, technological mediation may enhance an epistemological relation to absent Others and allow for an empathic connection with Others who are merely technologically present.

Despite stunning methodological innovations initiated by Husserl, whose *Wissenschaft* program besides other things was designed to understand and bridge various scientific disciplines, phenomenologists have not paid much attention to medicine except its psychotherapeutic applications (cf. Spiegelberg 1972), and these usually took a Heideggerian turn in exploring the so-called "humane" side of medicine as the investigation of patient's modes of being (L. Binswanger, V. Frankl, M. Boss, R. May, E. Spinelli, R. Zaner, and others). All this only mirrored and even amplified the major confrontation between "biomedical-empirical" and "humane" approaches in medical science and praxis, which in my view calls for a serious reevaluation. Another outcome of this mainstream approach was the forgetfulness about the physician, or in phenomenological terms, the loss of the whole field of transcendental performances as the conditions for the constitution of diagnostic judgment.

I. CATEGORIAL INTUITION AND EIDETIC VARIATION IN DIAGNOSTIC RADIOLOGY

In scientific imaging, the relation to the image is strictly functional, because the task of the observer is to extract nonrandom patterns from image data, or speaking in terms of physics, to look for low entropy states, like biological systems, "highly ordered and requiring a lot of energy for self-maintenance" (Bryan 2010, 84). This indicates that despite sophisticated physics behind it, steps of *medical imaging* and relevant diagnostic analysis are rather simple. The goal is a clinical decision based on the image, focusing on the presence or absence of cancer. The chain of information transmission consists of the organ's tissue, the imaging system, various modes of imaging and the observer, for example, human and/or computer, and the eye, that is, trained tendencies highlighted by tracking experiments which reveal visual preferences in identifying a new pattern (pattern recognition) or in comparison to previously known patterns (pattern matching or classification). Finally, when a match is made, a decision is made.[1] It could take only five seconds for an experienced radiologist to detect abnormality in a chest Xray. Yet, despite its straightforwardness and simplicity, the process is phenomenologically complicated.

With the "empirical approach" in medicine, we encounter what Husserl called the natural (or naïve) attitude (Husserl 2006).[2] It is rooted in (1) the unexamined belief in objects and patterns "outside" experience; (2) usually a "passive" understanding of consciousness; (3) a clinical judgment as a kind of opinion-based act; and (4) a simplifying tendency to explain the process of imaging by perceptual consciousness alone (or at best by relating cognition and memory to specific visual tasks). Yet, even natural scientists admit crucial gaps in naturalistic explanations of imaging.[3] In fact, despite Husserl's multiple approaches to phenomenology (cf. Scalambrino 2015b), already in his *Logical Investigations* (1901/2001), Husserl was concerned with the *Noema* of intended objects *presented in their intuitive absence*, as in the cases of memory, expectation, or with a bit of difference in imaging. Moreover, the phenomenological notion of "intentionality" presupposes that everything that is meaningfully given to consciousness must be structured, that is, modified before its immediate givenness in perception.

An image, then, as the depiction of a state of affairs (*Sachverchalte*) always involves a *categorial intuition* grounded in passive synthesis (*sense intuition*). *Categorial intuition* is presented as soon as something is understood *as* something, and diagnostic judgments always exhibit the form of a categorial act. For example, a tumor can be detected *in* the lung, brain, hand, etc. Notice, though here we have any number of elements which can be easily fulfilled in sense perception (perceiving a hand, a brain, etc.), but what in experience gives fulfillment to the "being contained in something or relevant to something else" of a tumor?

The basic difference between "levels" of intuitions is that simple intuition in the form of sense perception presents its object "directly," that is, "immediately," and its presenting function does not rest on founding acts (*Fundierung*), while *categorial intuition* is founded. But the most important moment, which opens the way to the transcendental level and permeates all phenomenology, is that *Fundierung* have no "actual" properties such as causality, logicality, or psychic associations. These relations are nothing more (or less) than "meaningful" (Crowell 2006/2009, 14). Further, *Fundierung* are followed by another correlative phenomenological relation among intentional acts, that is, "fulfillment" (*Erfüllung*). For example, in the case of judgment "this tumor is malignant." it is not that the judgment rests on "its" perceptual content; rather, it is simultaneously *fulfilled* by it. Judgment and the articulation of perceptual content constitute purposeful, or teleological, thinking, but not the causal order.

In *Logical Investigations*, Husserl gives many examples in regard to forms of *categorial intuition*, but one of them is of special importance for our topic. It is the celebrated *eidetic reduction*. The purpose of the eidetic reduction is to strip all the individual features of the "object" and shift attention to the "essential" ones.[4] For example, a particular shape can be the object on which a radiologist is focusing when her eyes are directed toward an image of tissue. The method by which the reduction is achieved is eidetic variation,[5] which calls for the imaginative and systematic variation of examples under study while features that constitute abnormal *Sachverhalte* emerge.

II. MEDICAL IMAGING AND IMAGE CONSCIOUSNESS

Following phenomenologists such as Aron Gurwitsch and Husserl's later transcendental disclosure, then, that the structure of experiencing states of affairs, *Sachverhalte*, requires some means beyond sense perception (cf. Scalambrino 2013b), *we encounter a fundamental need for "mediating" modes of consciousness* as the ground for the possibility of any theoretical operation. Thus the super-sensible or *categorial intuition* formulated in *Logical Investigations* is followed by the theory of imagination:

> Consciousness of what is *not present* belongs to the essence of phantasy. We live in a present; we have a perceptual field of regard. In addition, however, we have appearances that present something not present lying entirely outside this field of regard. (Husserl 2005, 63)

Moreover, in the course of the lectures collected in *Husserliana XXIII* (Husserl 2005), Husserl differentiates Perception (*Perzeption*), Image or Pictorial Consciousness (*Bildbewusstsein*), and *Phantasie* (imagination proper or immanent imagination). Hence, already basic differences are found in the distinction between presenting and presentiating consciousness (*Vergegenwärtigung*,[6] characterized by reproduction, modification). That is to say, perception is the form of presentation (positions objects as present, *in propria persona*), while *image consciousness* and *Phantasie* account for the acts of presentiation, that is, presenting objects "as if" they were real, actual, or present and presentiating as nonreal, nonactual, absent, Other, respectively.

In the case of a medical image, clearly the experiential structure of *image consciousness*[7] is involved, which with its physical media (radiological images) remains rooted (founded) in perception but uniquely (i.e., neither *Perzeption* nor *Phantasie*) exhibits the structure of a twofold apprehension (when the appearing image and the "subject" of the image do not coincide). Thus, the elements involved in *image consciousness* include the following:

(1) An actual physical thing—photograph produced by the particular mode of imaging, that is, the physical and sensible basis for the image
(2) The image-object (*Bildobject*)—image, depiction of the system of organs, that is, what visually appears to spectator
(3) The subject of the image (*Image Sujet*)—what is *meant* or intended, for example, state of functional organs or other body parts

Without phenomenological reflection, it is supposed to be the case that (1) and (2) do not stand for themselves, but rather open the way for a *consciousness* of (3). The perceiver perceives the *physical substrate* as an actuality, but the *image* (system of organs) is not experienced as an actual existent; however, the *subject* (the system of organs depicted) might be so experienced though nonpresent. This means that the *image consciousness* does not include the belief or, more precisely, the *thetic* characteristic proper to perception wherein the perceiver takes the "object" as an actual existent here and now before her. We may now see how these three structural "stages" of *image consciousness* are echoed in medical imaging.

At the first stage of radiological analysis, a radiologist must evaluate the quality of an image, that is, shift her attention to the media as a physical thing, evaluating the degree of "noise," which interrupts signal. This evaluation can be improved technically, that is, through increased technological mediation, especially by mastering diagnostic monitors for the sake of zooming and altering the visual presentation's size. Moreover, a radiologist must consider from the outset such peculiar limitations pertaining to the concrete modality of imaging (Xrays, ultrasound, magnetic resonance, computed tomography, nuclear medicine, etc.).

Like an art student approaching a drawing by Picasso, a radiologist concentrates mostly on the second structural element noted above, that is, the *image-object*. Importantly, this is the moment where the interplay between *eidetic variation* and *anticipation* occurs. For Husserl, anticipation is an essential feature of human action; in regard to intentional action, then, the act is perpetually "empty" and heads for a realization that is determined by the *expected* essential features of normal (or abnormal) states of human body. Finally, *differential conclusion* indicates the last step of diagnosis concerning pathological changes and possible progression of disease, and this brings us to the following discussion of judgment and the third structural element of image consciousness noted above, that is, the *imaged object*.

III. PERCEIVER AS EVALUATOR: CONDITIONS FOR THE IMPROVEMENT OF DIAGNOSIS

Judgment (*Urteil*) is one of Husserl's favorite examples of *active synthesis*.[8] This kind of intentional act is at once aimed at the *categorial object* (*Sachverhalte*), that is, a particular situation or state of affairs (in our case, organs or bodily functions), and is structured according to one or another categorial form (such constitutional relations as "in," "under," "and," "near," "one," "many," "if," "not," "some," etc.): "Articulation of its perceptual content is the *telos* of judgment, the measure of its success or failure" (Crowell 2006/2009, 14).[9] Experiential structures, which are at play in the diagnostic process performed by a radiologist and which I deliberately summarized as "attentionality," can be approached via various phenomenological projects, for example, Gurwitsch's so-called "Field Theory" (1964).[10]

According to Gurwitsch, every conscious[11] moment exhibits this triadic structure:

a. *Theme*—focus of attention;
b. *Thematic field*—all phenomena appresented or copresent along with the theme in focus;
c. *Margin*—objects that are presented but at the same time are not endowed with the relevance to the dominating theme (e.g., features of embodiment, perceptual surroundings, images, affection, etc.).[12]

This structure, according to Gurwitsch, is *universal a priori* for every experiential building block in any mode of their givenness.[13] During the initial steps of the diagnostic process, a radiologist fragments the wholeness of the visual field into building blocks and focuses her attention on the areas that are more likely to have abnormalities, in the radiological sense. Radiologists are trained not only to reduce their evaluative gaze to relevant subareas but also to identify and "mute down" irrelevant information which, according to Gurwitsch's schema, would mean the ability to define precisely the margins of the thematic field. Those instances of irrelevant information (e.g., dust on radiogram, marks left by the static charge irrelevant for diagnosis, parts of human body, etc.) are called *artifacts* and are especially tricky for novice radiologists. Structural analysis of the thematic field peculiar to radiology now brings us back to the phenomenological account of perception because in the context of medical diagnostics it is appropriate to ask: *Under what conditions may a perceiver be counted as a diagnostic evaluator*?

It is a truism that from the phenomenological point of view perception is not just a passive reception of sensual data via some causally determined apparatus; rather it is active restructuring of the whole. The flow of perceptual data is not determined physically but depends on us via structural changes that we *are able* to make.[14] From the phenomenological point of view, a radiologist does not concern us as a psychological, positional ego; thus we must emphasize that the *subject*-ive correlate of the thematic field is an *attitude*, and it stands for how the theme appears to consciousness. Margins also (though only by negation) determine the thematic field, that is, their role is also crucial. Indeed, it is quite important that the distinction between the thematic field and the margin can be difficult to define. In other words, elements from the margins can penetrate the thematic field, thus shifting its significance.

The notion of thematic attitude, or perspective, is of a crucial importance regarding the evaluative abilities of a radiologist. If the perspective is loose, an evaluator will have some difficulties deciding what should count as relevant. The "break down" of a diagnostic process can also occur through prolonged concentration on a particular fragment, that is, "staring." In fact, though beyond the scope of this chapter, accounting for the possible mistakes in diagnostic judgment falls into several categories. Generally speaking, they may exhibit a *synchronic aspect* when simultaneous features of the surrounding spatial context confuse the evaluation or a *diachronic aspect* when the evaluator fails to use previously acquired information to respond to what should be a familiar context.[15]

IV. INTERSUBJECTIVITY AND EMPATHY BETWEEN "ABSENT BODIES"

It is peculiar solely to radiology to encounter a person not via "everyday" pictures but as a categorial constitution of inner-body space. In some sense, a radiologist serves as an amplification of this bringing closer or invitation of an Other *Lieb* based on angle or perspective or part of the radiological image.[16] Hence, a radiologist not only gets some knowledge about an *imaged subject* (that which is depicted) via the physical substrate of radiogram and the *image subject* (that which is intended), but is also displaced into the midst of *intersubjective constitution.* Now, we are ready to handle the question: *How precisely is a person given to a radiologist via medical imaging*? The quickest answer could be that the person is *quasi-given* to a radiologist, but what does that mean? Quasi-givenness is situated in quasi-perception ("perception" awakened by images), which exhibits a structure of repetition, that is, across repeated looking at different images it may be determined that the images refer to the same person.[17] The problem is that in radiology this quasi-perceptual repetition concerns the *inner*, or phenomenologically more accurate, *absent* (nonpresent but presentified) body of an-other person.[18] Thus, we must look for a more complicated structure.

According to Crowell, "phenomenological analysis shows that constitution of the perceptual world entails more than vision conceived as a mental act; it requires an *embodied* subject" (Crowell 2006/2009, 25). The importance of this notion of incarnation or embodiment consists in that it points toward the active role that the motile body, as the organ of sense, plays in the constitution of the objects and the world. But this body must be rediscovered. For in the natural attitude there are no Others because there is no me as an ego—all are the same real entities, "there are only real men" (Ricoeur 2007, 120). In his Fifth Meditation, Husserl shows how just after the reduction to the sphere of owness (*Eigenheit*), when I suspend everything alien, incarnation is reached as *Lieb*, that is, my body as an animated organism. This "additional reduction," as Ricoeur puts it, discloses physicality of my ego and motivates[19] the constitution of the Other. But the Other is not presented directly, just her body, therefore Husserl finds another "*mediating* intentional act [emphasis added]"—where objects are copresent—and this intentionality is called *appresentation*.

Appresentation as the "higher level" (categorial) order is grounded in the level of perception, more specifically, in the pre-reflective originary form of passive synthesis called *pairing* (*Paarung*).[20] Pairing in turn is a passive synthesis in which the members of the pair are intended together, by simultaneously triggering the sense of each other (e.g., "hammer" and "nails"). In the Husserlian understanding of the constitution of the Other, two such pairs go simultaneously:[21]

> *Pair of association*—when "I" associate my already noticed (reduction to the sphere of owness) body with the ego and thus constitute the *noema* "subjectivity."

> *Pair of similarity*—when "I" pair my body with the similar body (given by sense intuition) and thus constitute the *noema* "other subjectivity."[22]

It looks like for Husserl *pair of association* founds *pair of similarity* because "after" the association (in similarity) you would not expect a zombie or a cyborg in the *face* of the Other. This is to say that Husserl's passive synthesis, which requires further mediation, may perhaps be enhanced by technological mediation. Further, if these two pairings happen simultaneously, we can look in reverse: what if, in special cases, for example, diagnostic judgments based on medical imaging, similarity founds association? Hence, medical images afford a special power to "motivate" a radiologist in the direction of a specific constitution and doxic acts (e.g., judgment and expectation).

Whereas bodily images constitute a radiologist's lived space, in a sense (intentionally) a radiologist becomes so sensitive to the similarity of the Other body, as to be constantly provoked to perform the *association*, that constitutes *Lieb*. This empathic relation motivates the radiologist first to detect *Körper* in *Lieb* (an *abnormal* organ in an Other's body appresented as mine), and second to emphatically construct a narrative concerning the patient's healthy and diseased case history, which in technical terms turns out to be the differential conclusion (i.e., the diagnostic judgment).

Thus, perhaps more importantly, medical images become that sort of media, which help to improve empathy, which in turn become the culmination of knowledge through and in medicine.[23] Therefore, our experience of perceptual objects is constantly accompanied by pre-reflective experience of the position and movement of the body, named *kinesthetic experience.* This is like a movement from the basic phenomenological tenet that experience is always an experience of something to Merleau-Ponty's extension stressing that an appearance is always an appearance of something for someone (incarnated). That is to say, the structure of perception coincides with the structure of the body. More generally, as we have already seen in Husserl, the givenness of something in perception is never just the effect of an external cause; rather, it is the participation of the organism in a meaningful and multileveled situation.

One of the conditions, then, of every situation (*Sachverhalte*) is the physical surroundings, which in the case of radiological judgment are determined by specific technological media (radiograms depicting internal or absent categorial objects). These "surroundings" in turn afford, provoke, or awaken the intentional activity of imagination and empathy for a radiologist. Thus, first of all, corporeality sets up a *space* within which the transcendental ego exercises all its intentionalities. In other words, a radiologist pre-reflectively takes a position concerning the monitors, and visually prefers one or another object contrary to the whole horizon presented by radiograms. This enables a radiologist to attain what Merleau-Ponty calls "optimal bodily attitude," necessary for the constitution of the objects involved. On the one hand, though there is always a horizon of vision, in the case of radiological images this horizon is already narrowed down to the particular objects. On the other hand, the pre-thematic background, the unstructured whole of the subordination between particular organs and/or bodily functions, should guarantee the identification of the particular object in the course of diagnostic exploration.

When a radiologist looks at body parts, organs, and/or functions these are given to him in conjunction with the sensation of his own body, but it is not only the kinesthetic experience that serves as the condition of adequate perception, it is also the *imaginative displacement* into the realm of absent or inner embodiment that grants the radiologist the co-intention of directly absent profiles of medical "objects" and possibilities for diagnosis. The absent profiles are linked into an intentional if-then connection. Thus, judgment is also presupposed by bodily awareness and the horizon of imaginative possibilities.

This *imaginative displacement* is possible for a radiologist only on the ground of the self-experience of her own inner-body constitution and deep empathic affinity with another body which is still absent *in propria persona*, while already presentified by radiological images. Thus, while identifying diagnostic objects, a radiologist also continuously identifies her own body as the privileged space, nurtured by medical training (briefly discussed above), and in relation with her surroundings of medical images (technological media) manifests a place for the *absent* human body *to show itself*.

V. CONCLUSIONS

Phenomenological analysis of the methods of empirical diagnostics enables us to deconstruct the gap between "humane" and "empirical" approaches to medicine. On the other hand, the constitution of diagnostic judgments based on medical imaging presupposes a very specific intentional analysis comparable to the face-to-face encounter between a physician and her patient. A number of Husserlian concepts (*categorial intuition*, *image consciousness*, *empathy*, etc.) were fruitfully applied for navigating the concrete stages and technologically mediated tasks of diagnostics based on medical imaging, simultaneously disclosing transcendental conditions of this empirical-inferential process. The results of this descriptive analysis might be used for the improvement of the perception, analysis, and evaluation of images involved in the diagnostic process. Hence, Husserlian phenomenology may disclose how humans already merge, through imagination and empathy, with technological mediation toward the public good of diagnostic judgment.

NOTES

1. Inability to perform such cognitive matching skills functionally amounts to neurological disorder, for example, *agnosia*—"fundamental lack of recognition" (Carter 2010, 188) pertaining to the "objects" of various kinds.

2. Analogically, in the "humane" approach to medicine, we encounter another modification of this naïveté, that is, the psychological ego (cf. Husserl 2006).

3. First, it is not clear how an individual chooses a particular approach to a specific task, for example, that of pattern localization or classification. And "higher-level" tasks, such as identification of complex shapes like cancerous tumors, are not well understood. Secondly, the central dilemma for neuroscience, the so-called *binding problem*, is this: How are the outputs of these disparate channels (i.e., individual features of the "object") recombined into a single perception?

4. P. Ricœur rightly notes that not every shift of attention is worth the status of "reduction" (2007, 118–23).

5. Husserl also refers to this method as "imaginative variation" or "free phantasy," which he first borrowed from the philosopher and mathematician Bernard Bolzano's *Theory of Science* (1837/1972).

6. This term can be translated as *presentiation* or *presentification*, but clearly not as *re-presentation*: "On no account should we fall into the fundamentally perverse copy and sign theories . . ." (*Ideas 1*, § 52).

7. *Phantasie* is not operable while *perceiving* pictures.

8. In the wake of Immanuel Kant, Husserl distinguishes between the understanding of "judgment" as *Urteil* (a kind of "cognition") and *Satz* ("proposition"). This "double character" of judgment that is its ontological factuality of categorical form and logical factuality of proposition must be acknowledged. However, only through phenomenological reflection can we identify logical meaning (*Bedeutung*) also as phenomenological sense (*Sinn*).

9. In phenomenology, neither objective temporality nor causality can explain relations like this—their genesis has its own logic guided by the noematic structure, thus this is a sense-based relation, not the causal one.

10. Cf. A. Gurwitsch, *Field of Consciousness* (Pittsburgh, PA: Duquesne University Press, 1964).

11. Here, "conscious" is used in a phenomenological sense, that is, meaning *every* act of consciousness, contrary to common understanding of "conscious" as self-reflective act. See in this regard, Zahavi, D. 2001), *Subjectivity and Selfhood: Investigating the First-Person Perspective*, Massachusetts Institute of Technology.

12. It must be stressed that this tripartite structure is not stable at all and we can easily be forced to shift our attention to the things which just a moment ago were marginal because of a new visual data, changing spatial context or inflicted pain.

13. Perhaps the fundamental intentionality of time-consciousness may also be mentioned here.

14. Cf. M. Merleau-Ponty, *Phenomenology of Perception*, trans. D. A. Landes (New York: Routledge, 2012).

15. Cf. L. Embree, "Gurwitsch's Relevancy for Cognitive Science," *Contributions to Phenomenology,* vol. 52 (Springer 2004).

16. Another important thing concerning the pictorial intending and its affinity with perception is its repetitive nature—we can focus on one or another part of the image, get back to unclear parts, etc.

17. Cf. E. Husserl, *Collected Works: Phantasy, Image Consciousness, and Memory (1898–1925),* trans. John B. Brough (Springer, 2005), 656.

18. Though I took the concept of "absent body" from D. Leder's book, here it enters into different phenomenological projects than those analyzed by the author of the concept. See D. Leder, *The Absent Body* (Chicago: University of Chicago Press, 1990).

19. This motivation by itself constitutes an independent and intriguing question, the answer to which lies in examination of the urge to the objective (intersubjective) world or reality.

20. Ones again we see a meaningful and not causal relation between simple and categorical intuitions.

21. In the phenomenological tradition, there are at least three more approaches concerning the problem of the constitution of the Other (cf. Zahavi 2001).

22. Because these pairs are only implicit in Husserl's text, I use K. Hermberg's division (Hermberg 2006).

23. I argue that this lived space fostered by medical images grounds the deep empathic relation between radiologist and other ego, and perhaps it should be better called "bodily empathy" because even the *proprioception* or constitution of the inner space and location of my organs (in the absence of any special technological mediation or sensory organ for this experience) is involved.

Chapter 15

Absent to Those Present

The Conflict between Connectivity and Communion

Chad Engelland

Social technology allows us to transcend where we are and connect to a virtual community of friends wherever they happen to be. It is not uncommon, for example, to be sitting around a table with people who, thanks to their devices, are silently connecting to those who are not there. Bodies are in one place and minds are in another. The ancient Greeks had a saying about daydreamers like these: "Absent while present" (Heraclitus 1979, 29). Today, social technology makes us absent to those present in two ways: first, our minds are absent to those around us and second, our bodies are absent to those we are connected to. Either our minds or our bodies are absent. What social technology cannot deliver (and in fact encourages us to ignore) is intentional bodily presence. Should we be concerned by this absence? I think so, because communion begins in and is nourished by bodily presence, and we humans have an innate need for such communion.

In this chapter, I first analyze the technologically mediated connection, which I call "connectivity," and contrast it with other ways of communing with family members, neighbors, coworkers, and friends, which I refer to simply as "communion." I argue that connectivity competes with communion, because such connections make it possible to be always present to those who are absent, which correspondingly makes us absent to those who are present. I provide a phenomenological description of immediate and mediate interpersonal presence (in person, by telephone, by video call, by letter, by email, by text, and by social media) to show how and why recent technology differs from older forms such as telephone and letter writing. I then call attention to the importance of bodily presence for affording communion. Finally, I point to the importance of absence and thoughtfulness for giving us something real to talk about when we are together. Interpersonal presence draws its life from our bodies and our thoughts.

THE NOVELTY OF CONNECTIVITY

We say that our devices allow us to be "connected" in new and exciting ways that people have never experienced before. For example, in 2012 Google CEO Eric Schmidt gave Boston University's commencement address, saying:

> You're connecting to each other in ways those who came before you could never dream of. And you're using those connections to strengthen the invisible ties that hold humanity together, and to deepen our understanding of the world around us. You are emblems of the sense of possibility that will define this new age. (Schmidt 2012)

What is a "connection," and how is it different from the way people used to relate to one another before the advent of social technology?

Connections rest on the temporary choices of the person. The individual effortlessly chooses friends, products, and services. One week we may connect with so and so, the next week we may not. Such choices are not even constrained by geographical proximity. We can connect to people across the globe in Australia as easy as people across the room. It doesn't matter where our body is, for our minds can meet in cyberspace. In communion, by contrast, daily interactions were dictated by proximity and bodily presence. People had no choice but to commune with the family members with whom they lived, their neighbors, their coworkers or schoolmates around them, and nearby friends. Unlike connectivity, then, nontechnological interactions related to shared situations, such as home, work, or school. They were for the most part assigned to people instead of being chosen by them. The number of interactions was determined by the effort of bodily presence needed to bring them about.

The difference between communion and connectivity, then, is that communion presupposes a community of people to which one belongs without specifically having chosen to do so. It relates to a whole pattern of life built around natural needs for food, shelter, community, and understanding. Connectivity, by contrast, presupposes no such community and instead makes the relation turn on the effortless preferences of the individual. Our devices free us to transcend bodily presence to establish virtual presence to a multitude of absent people.

What is the relation between connectivity and communion? Doesn't connectivity just build on top of communion, as a mere extension of friendship? Precisely because connectivity denies proximity and bodily presence, it is always available (save for power failures, battery outages, and network glitches). Moreover, when we are connecting we are not communing with those around us. As we travel through the world, driving, walking, working, shopping, eating, and so on, the people around us appear insubstantial compared to the continual conversation connectivity always affords. When we are bodily present to others we are always on the verge of leaving them for connectivity (always looking for the least provocation to check our devices), and when we are connecting, we are oblivious to those around us. Connectivity crowds out opportunities for community with family and friends. Family members need no longer interact with one another, because connections are always available. Friends present to one another do the same thing they do when they are absent: connect to absent people. To be present to those bodily absent, we acquired mobile devices with the unintended consequence of being absent to those with whom we are bodily present.[1]

DEGREES OF PERSONAL PRESENCE

Perhaps I am being too hard on connectivity. After all, didn't letter writing and the telephone already allow people to choose who they wanted to interact with? Language allows us to talk about things that are absent (I can tell someone about a trip abroad), and technology enables us to present language in the absence of the language speaker (you're reading my words right now although I'm nowhere to be seen). Technology affords conversation across the distance separating bodies in two ways: by conveying the *sound* of our speech (telephone and video call) or by conveying our *written* words (letters, emails, texts, and social networking). Today, the connectivity of email, texting, and social networking encourages continual, instantaneous production and consumption of content by an increasing number of connections. As I will show in this section, they thus differ from older forms of technologically mediated social interaction such as letter writing and phone conversations.

In Person: Among all the ways to converse, conversation in the flesh is the most powerful. Unfolding in real time, supported by the natural expressiveness of our bodies, such conversation allows us to encounter another person and, in doing so, to come to know ourselves. Conversations occur naturally in the context of a focal practice or joint activity such as going for a walk, having a meal, and drinking coffee.[2] Buoyed by the sunny day or jointly affected by each other's sorrowful moods, our communing occurs not only in conversation, but in a wider setting afforded by the fact that our bodily natures are present to each other. The tasty meal (or the overcooked one) affords a preconversational communion out of which our conversation naturally arises. We can face the faces of other people, see them seeing us, or respond to their responding to our response: "When we make eye contact, our attentions join and we face the world together. Here we do not look *at* the other's eye, but *with* them. . . . Such a look says, 'I see you seeing me' or 'you see me seeing you' and 'together we are considering the same thing'" (Engelland 2010, 451). This ability to face each other supports the back-and-forth, give-and-take characteristic of conversation. It enables us to see how another is affected by what we say and such affection makes us vulnerable and bonds us together. There simply is no substitute for the raw exposure of self to self that happens in the flesh. Naturally, we can still undermine interpersonal communion even in this context. For example, St. Augustine went so far as to carve a rule prohibiting gossip into his dining room table: "Who injures the name of an absent friend / May not at this table as guest attend" (Possidius 1919, 95). The table is a focal point for fellowship and bodily presence.

By Telephone: Alexander Graham Bell's invention allows conversation to occur across distance at the same time even with those sitting at other tables. It does so by conveying the presence of one's voice alone. The body is visually absent but the telephone allows the tempo and tone of the voice to carry much of the mood of the other. There is no visual object of our attention, which means we can become wholly absorbed by the conversation, not tethered by any perceptual support from our immediate surroundings. While on the phone, we are consequently absent to those about us. The advent of the mobile phone transforms the telephone. Instead of people calling someone's house, they call someone. Before we might reasonably not have been at home; now, however, we are always available to a would-be caller and that also means we are always ready to be distracted from whatever we are doing or whatever conversation we are presently having.

By Video Call: The video call adds to the telephone a presentation of our interlocutor's body, her facial expressions, her gestures, as well as the setting in which she finds herself. It conveys more of the mood of the other person, but it also serves to heighten the jarring difference of setting. We peer out from our vantage point into a very different place. The video message calls attention to the fact that we are not in the same place in a way that the phone call passes over. Unlike a conversation in person, there is no joint domain of mood and action. We aren't necessarily under the same weather, the same time of day, and it is more difficult to gage and receive the mood of the other. We cannot share the same pot of coffee, take a stroll, or enjoy the same meal together. Also, due to the mediation, it is impossible to look into the other person's eyes. Either we look into the off-center camera, so that they see us looking directly at them (with the tradeoff that we are not directly looking at them on the screen) or we look directly at them (with the tradeoff that our eyes do not look into the camera at them but away from them at the screen).[3] The result is that in a video call each appears distracted from the other. We can see a moving picture of the other person, but we can't *see them seeing us*, the dynamic that makes conversations in the flesh so communicative.

By Letter: The art of writing a letter allows conversations to occur across distance although not at the same time. Handwriting betrays something of the writer's person and mood. The introduction of the typewriter and then the keyboard removes these vestiges of the body. The words themselves could be anybody's. The signature at the end is the last vestige of authenticity. In letter writing, the form of expression has the virtue of inviting reflection on just what will be expressed, since there is some effort in expression; consequently, there is a tendency to avoid idle gossip and to speak of more important things. Yet what is lost is much of the back-and-forth interplay as it occurs in the midst of conversation in the flesh. And there is no joint setting or joint activity.

By Email: Email communication stands between letter writing and texting. The signature, the last vestige of the body in a typed letter, gives way to the email address as certifier of authenticity. The email adds the possibility of effortlessly copying a message to many. It is ideally suited for informing, not communing. Email messages with the content of a letter are strange and frankly overwhelming; as welcome as they might be we don't really want to go through the trouble of responding to them. It is too much trouble to write something substantial; it is much easier to forward a message or send quick little bits of information. Due to its ease, emailing becomes much more frequent than letter writing. It also differs in this: whereas mail comes but once or twice a day, emails arrive episodically but frequently throughout the day. Email encourages an increase of messages with a corresponding decrease in significance.

By Texting: Texting is a form of writing that allows conversations to take place across distance at the same time. There is a back-and-forth interplay lacking in letter writing and emailing, but the short format and instantaneous conveyance that makes this possible further discourages thoughtfulness and encourages quick, short replies. A letter does not demand to be answered immediately, because it takes time to be delivered, but texts and calls, since they are virtually instantaneous, demand immediate answering. The tradeoff between quantity and quality, introduced by the email, becomes heightened (Turkle 2012). Emoticons and emoji try to substitute for the lack of personal expression afforded by the anonymous means of communication. The quantity of texts and shared images increases so that we constantly receive new ones.

By Social Media: Social networking allows us to keep track of others and communicate with others via mediation of some kind. We post a message or a picture that we make available to our connections. In such a situation, we do not so much talk to others and are not so much talked to by others as we talk about ourselves and hear other people talk about themselves. We learn much *about* many others but we do not *encounter* them (Dreyfus 2001/2009, 49–88). Selfies testify to a deep desire to be encountered. Sharing a selfie says, in effect, "Look at me!"[4] However, the image does not allow *us* to be seen: my viewing your image and your viewing my image does not capture the dynamic interplay of personal presence. In the flesh, you see me in the act of seeing you and I see you in the act of seeing me. In social media, our appearance is truncated. Also, we have to keep our messages short because they are being broadcast to so many and because our readers, consuming so many different posts, do not have the leisure to make their way through something longer. We also do not have much to say, since our own habit of communication and thus thinking has been fashioned by the short bursts of texts and images that fill social media. Social media always provides us with something productive, stimulating, and easy to do: read our feeds and occasionally make comments. We are informed, and we are entertained.

INTIMACY AND BODILY PRESENCE

Technology allows us to publish our thoughts and be more widely known at least concerning those bits of ourselves that can be conveyed in this way: facts, fancies, and photographs. But the more it makes us publicly known the less it makes us intimately known. Technology allows us to connect with people who are absent by leaving aside our full bodily presence, but it is just this full bodily presence that affords the possibility of experiencing true intimacy and interpersonal communion. Our bodies, moods, shared activities, and eye contact contribute much to the communion of human persons (Engelland 2014, 131–70). Intimacy requires exclusivity, which requires our living bodies. Augustine spoke of the many bodily acts that solidify a friendship:

> To talk and to laugh with them; to do friendly acts of service for one another; to read well-written books together; sometimes to tell jokes and sometimes to be serious; to disagree at times, but without hard feelings, just as a man does with himself; and to keep our many discussions pleasant by the very rarity of such differences; to teach things to the others and to learn from them; to long impatiently for those who were absent, and to receive with joy those joining us. These and similar expressions, proceeding from the hearts of those who loved and repaid their comrades' love, by way of countenance, tongue, eyes, and a

> thousand pleasing gestures, were like fuel to set our minds ablaze and to make but one out of many. (Augustine 1960, 4.8.13)

Some of these activities can occur via social technology, but even then they lose some of their allure because of the absence of the body. We can read a joke on a Facebook wall and laugh, but it is funnier when we are bodily present and can react to the bodily reactions of others—that's when laughter rages out of control and causes us to split our sides. Intimacy comes not only from the content of our conversation but also from the reciprocity of our bodily bearing that confirms the other person even as one is confirmed oneself: "Intimacy is rooted . . . in the unique *being present* of a person; and the principal mark of intimacy is *attentiveness to the presence of another*" (Schmitz 2007, 161). The difference between connectivity and communion comes out most forcefully in those moments when people vainly try to establish communion with a connection; consider, for example, the case of sexting in which people try to be intimate with a connection only to expose themselves to others.[5]

There is no natural limit to how many connections one can have, because the individual connections require little bodily effort. At the same time, these connections do not satisfy our craving for friendship and intimacy, because the quantity and quality of mediated conversations are inversely related. Aristotle thinks we should want as many close friends as possible, but, since close friendship requires bodily presence and focal practices, it is not possible to have many: "As to our seeking and praying for many friends, while we say that the man who has many friends has no friend, both are correct. For if it is possible to live with and share the perceptions of many at the same time, it is most desirable that these should be as numerous as possible; but since this is most difficult, the activity of joint perception must exist among fewer" (1984, 7.12, 1245b20–24). Intimates share a table together and tables can only be so big. The body enables a rich presence, its absence a poorer one, but precisely because it is poorer it can be multiplied without effort thanks to technology.

ABSENCE AND THOUGHTFULNESS

So far I have highlighted the importance of bodily presence, but now let me say something surprising: absence is important, too, because bodily presence is not enough for communion. We have to have something to talk about. Here thoughtfulness is crucial, and thoughtfulness requires solitude. Henry David Thoreau observes, "When our life ceases to be inward and private, conversation degenerates into mere gossip" (1993, 360). He also realizes that the absence of inwardness leads us to seek more earnestly a quantity of conversations: "In proportion as our inward life fails, we go more constantly and desperately to the post office. You may depend on it, that the poor fellow who walks away with the greatest number of letters, proud of his extensive correspondence, has not heard from himself this long while" (1993, 361). Proud as we are of our connections, there might be someone we're not connected to: ourselves. Since we are always present to others, we have no occasion to think for ourselves and therefore to know ourselves. And this means we have little to contribute to conversation when we are present to others. We can only repeat what we've heard, not what we've thought.

Presence is an achievement, and it requires absence and solitude to be properly nourished. The sort of presence achieved by returning from absence is the sort of presence that can reach deeper into the things of the world. People were desperate for Socrates' presence and yet, from time to time, he would wander off lost in contemplation (Plato 1997d, 175b). When he would return he said things that were startling, strange, and wonderful. He led people to see the world in a deeper way. He couldn't have done this if he spent all his time in the presence of others without himself drinking from the depths of things. Now, none of us is a Socrates, but we can always be more thoughtful than we are if we but withdraw from the buzz to think. In 1970, after my parents had been dating for nine months, my mom went with some friends to Hawaii for ten days; my dad felt her absence keenly and thoughtfully; when she returned, he proposed. Had their relationship begun today, they would have been connecting constantly and he would not have had much time or occasion to consider her absence and thus to appreciate her presence nor would he have realized the full depth of his own feelings for her. Full presence requires the presence of original thought, thought nourished by genuine contemplation, and the presence of the body. Incessant mindless presence in fact makes us absent to everybody, ourselves included. We do not have time to know ourselves since we know everything about everybody else, including what they just bought on Amazon and what they made for dinner.

BEING PRESENT TO THOSE PRESENT

If connectivity does not satisfy our deep desire for communion in the flesh then why do we continually sacrifice opportunities to commune for opportunities to connect? What is so alluring about connectivity? Why does it trump the more meaningful alternative? There is something deeply puzzling about connectivity and its mysterious hold on us. Perhaps part of the answer is the risk built into communion in the flesh. When we are with people, we have to rely on the inspiration of the moment; we blurt things out and cannot take it back; there's no delete button or opportunity to edit. Also, people look at us with eyes that not only love but also judge; we are vulnerable to the gazes of others. Communion in the flesh takes so much time and we are desirous to cut corners wherever we can for the sake of ease. If connectivity can give us conversation without costs it seems better. Moreover, we are deeply afraid we might be missing out on something happening elsewhere. We do not appreciate the ebb and flow of conversation; as soon as a lull presents itself in the flesh, we switch to connectivity. However, spontaneity, vulnerability, time, exclusivity, and the interplay of activity and rest are precisely those ingredients that make intimacy possible. Communion is a difficult but great good; if we turn from the difficulty to an easier substitute we should be mindful that we are also turning from something that will satisfy us to something that won't.

The sort of connections afforded by social networking can allow us to keep "in touch" with those whom we have already encountered in the flesh. But the tradeoff is that we are "out of touch" with those around us as well as our own bodily selves. Previous ways of communing with those absent were occasional, but social networking is continuous and persistent. As social networkers, then, we are always present to those absent and absent to those present. But it is being present in the flesh that really animates friendship, that begins it and that nourishes it. We humans are wired for communion—nothing else will satisfy us—and so we would do well to recollect, disconnect, and commune: (1) *recollect* by recalling our desire for communion and the inability of connectivity to deliver, (2) *disconnect* by fasting from devices especially when we are present to people, such as dinner and hanging out, and (3) *commune* by doing things and being present to those we care about. Rather than living in the constancy of mediated presence, we should cultivate an appreciation for the interplay of presence and absence and a commitment to the primacy of bodily presence. Our bodies limit us to the here and now, but the here and now we share with others thanks to our bodies is no limitation; it is the continuous possibility for intimacy and communion.[6]

NOTES

1. Several psychological studies support the same point. See, for example, Pryzbylski and Weinstein 2012. On the unintended effects of technology in general, see Kass 2002, 29–53.

2. On focal practices, see Borgmann 1984, 196–210. On eating as a focal practice, see Kass 1999.

3. Perhaps in the distant future, technology could overcome this problem by creating goggles that would allow us to face a three-dimensional hologram or avatar of our interlocutor; perhaps we would be able to look into the image of their eyes; but we still could not shake hands, hug, or share a meal. Such presence would still fall short of bodily presence.

4. I am grateful to Marty Dober for this observation.

5. I am grateful to Kevin Klonowski for this point.

6. For comments on earlier drafts of this chapter, I am thankful to Damian Ference, Marty Dober, Kevin Klonowski, Silvia Madrigal, Claudia Schussman, Greg Schussman, Camille Kennedy, and members of the University of Dallas philosophy club.

Chapter 16

Recognizing the Face and Facial Recognition

Levi Checketts

"Augmented reality" was once only an idea within the domain of science fiction. In the media of novels, short stories, movies, television shows, and videogames, writers could explore philosophical questions about the nature and effects of wearable, reality-mediating technology. Today, however, what was once mere speculation has become reality. Augmented reality is being achieved presently in the development of devices such as Google Glass. Although this technology is designed to be little more than a wearable smartphone, software developers have quickly realized new potentials for a technology that is always before the user. One case of this is the integration of facial recognition software, technology that already exists separately and is used by social media like Facebook, into augmented reality glasses so that the wearer is able to easily identify others.

Facial recognition software works by matching facial features and structures of a person's face with different photographs. The software then provides the user with information about the other person, including information listed on Facebook, LinkedIn, Twitter, dating websites, and even police databases. This allows the user to know the people around her, at least as well as they can be known through social media profiles, but this presents new questions that we must consider: How might social interaction mediated by facial recognition software affect the way in which we morally perceive other people? How might facial recognition affect our ethical response toward them? What implications does this technology have for social epistemology? These questions prompt us to carefully examine the way that facial recognition technology affects our perception and knowledge of moral action in the social realm. To better consider these questions, it is helpful to look at three different philosophies (Sartre, Marcel, and Levinas) of experiencing other persons, and analyze what facial recognition software suggests for them. Like many technologies, facial recognition has beneficial and harmful elements. While facial recognition advocates point out that such technology aids social interactions with others, the use of this technology challenges the possibility of encountering other persons as an Other and of genuine intersubjectivity, issues that demand careful attention.

I. THE CASE FOR FACIAL RECOGNITION

For the sake of this chapter, the facial recognition software I primarily consider is "NameTag," a new software application that is available for smart phones and Google Glass. The creator of the app states,

> It's much easier to meet interesting new people when we can simply look at someone, see their Facebook, review their LinkedIn page or maybe even see their dating site profile. Often we were interacting with people blindly or not interacting at all. NameTag on Google Glass can change all that. (NameTag 2015)

The argument, then, is that being able to see "who" someone is, through her social media profile, will allow us to better engage with her as who she is. Thus, rather than being in a coffee shop or subway car full of nameless strangers whom I would be unlikely to engage with due to my ignorance of who they are, facial recognition provides for me the possibility of engaging with someone whom I find interesting or attractive, leading perhaps to friendship or romantic involvement (or conversely, avoiding those who show up to me as uninteresting or unattractive).

Such technology also has enormous potential in benefiting society by identifying criminals,[1] by allowing people with facial recognition disabilities to better cope with the world, and even perhaps by making some forms of identity theft more difficult. There are also undoubtedly as of yet unrealized potentials for this technology, the same way that the internet has been used in things like protest action (through the work of groups like Anonymous) and research through "crowdsourcing" projects, acts which were unforeseeable in the early years of the internet.

For the sake of this chapter, I only give attention to the issue of "encountering others." Perhaps on the whole, encountering other persons is better than not encountering them, but the claim that facial recognition technology will facilitate this is far from obvious. We will know more *about* the other—that is true—but will we be more likely to know our moral obligation to the other, and will facial recognition facilitate ethical exchange between persons?

II. SARTRE AND THE OTHER AS OBJECT

In order to understand the work of Jean-Paul Sartre, it is first necessary to distinguish between object (being in-itself, *l'etre en-soi*) and subject (being for-itself, *l'etre pour-soi*). The object "is itself" while the subject is defined "as being what it is not and not being what it is" (Sartre 1984, 28). The object, the being in-itself, is essentially what it is and can be no other (Sartre 1984, 28ff). On the other hand, the subject, being for-itself, is defined primarily in terms of freedom, "the human being putting his own past out of play by secreting his own nothingness" (Sartre 1984, 60). In other words, objects are static, unchangeable, things-to-be-acted-upon, while subjects are dynamic, ever-changing, beings-which-act. Contrary to a subject whose freedom entails an ability to always be different, an object can be circumscribed and defined in terms of its properties.

This distinction is important because Sartre conceives of ordinary human interaction as a matter of subject-object encounter. In viewing the Other, I view him not as a co-subject or another like me, but rather as an object. Not only is it the case that "one of the modalities of the Other's presence to me is object-ness," but also inasmuch as I look at all upon the Other, I see him as object. In fact, I cannot even be sure of the Other as being a fellow subject because "whether or not [the Other's] consciousness exists in a separate state, the face which I see does not refer to it; it is not this consciousness which is the truth of the probable object which I perceive" (Sartre 1984, 340). Thus, I do not see the Other as a conscious subject like me. Though Sartre realizes that I can attribute to the Other different intentions and ideas, it is still the case that I apprehend him as an object insofar as he is not me and therefore not subject.

However, this phenomenon reverses when the Other looks at me. In "being-seen-by-another," I lose my subjectivity insofar as I become an object to her (Sartre 1984, 345). Just as my looking upon the Other makes her an object to my subject, so as she looks upon me, I find myself becoming object to her subject. This entire process of objectivation of the Other Sartre refers to as "the Look," which he illustrates in the example of one who is peering through a keyhole and hears footsteps behind her that suddenly stop (Sartre 1984, 347ff). The Look is manifest in the knowledge of being seen by the Other, and entails the shame of being made into an object by a voyeur (Sartre 1984, 349). In being made into such an object (*en-soi*), and especially in being made to feel shame in my object-ness, I am motivated to reassert my subjectivity by in turn objectifying the Other (Sartre 1984, 386). This is all to say that the process of being viewed as an object by the Other motivates a sort of battle of the wills wherein I must claim my own subjectivity at the cost of the Other's: "My reaction to my own alienation for the Other [is] expressed in my grasping the Other as an object" (Sartre 1984, 400).

The Other is therefore one whom I can never encounter except as opposition; "the essence of the relations between consciousnesses . . . is conflict" (Sartre 1984, 555). All relationships, including lovers and friends, boil down, for Sartre, to merely assertion of oneself over the Other, or being overcome oneself by the Other.[2] Thus, even the preliminary encounter with the Other reduces to the Look: "If I experience him with evidence [that is, being-seen], I fail to know him; if I know him, if I act upon him, I only reach his being-as-object and his probable existence in the midst of the world" (Sartre 1984, 400). In seeing the Other as object, I subject her to my will, in seeing her as subject, I subject myself to hers; in both cases, the encounter with the Other is one of violence and subjugation.

Sartre's understanding of the subject-object encounter never involves face-to-face meeting. On the surface, we might say the subject and object are facing each other, but neither one nor the other shows up *as* a face in the full human sense that I detail below. The other as object may show up as beautiful or ugly, as having blue eyes or brown, but never as being another subject. Conversely, to the object, the subject does not show up as face either: "The Other's look hides his eyes; he seems to go *in front of them*" (Sartre 1984, 346ff). The Look therefore relies on not seeing the other's face as a face. The other may only appear as a voyeur, that is, a subject who hides her eyes and thus her vulnerable face, or as an object, that is, a thing with features such as eyes, a nose, and mouth, but without a face like me.

In a striking way, facial recognition resembles the Look. In identifying "who" a person is, by matching facial features to pictures on social media sites, the software makes the object of view quite literally an object, that is, something determined and not self-determining. Furthermore, due to the fact that, in facial recognition technology, the voyeur has access to knowledge about the observed that the observed lacks, he is in a position where his subjectivity is unquestioned whereas the observed is presented as a describable being, not unlike a table or chair, or any other object that has properties and can be described. On the other side, the observed, when presented with an observer wearing augmented reality glasses, feels constantly objectified under the power of the Look; the Look can be accomplished through artificial means, such as the glaring light of a farmhouse at night (346). Augmented reality glasses, similar to glaring electric lights, thus create a situation of a "persistent Look," continually doing violence to the subjectivity of the Other.

III. MARCEL AND THE ENCOUNTER

If Sartre's view of human interaction is pessimistic, Gabriel Marcel's view can be noted for its optimism. Marcel elaborates a phenomenology of inter-subjectively encountering the Other. First, he notes, it is impossible for objects to encounter each other: "there cannot be an encounter or a meeting in the fullest sense of the word except between beings endowed with a certain inwardness" (Marcel 1950, 137). I cannot encounter an Other whom I perceive as object; I must perceive her as an Other: one who is able to be reflexive and "inward" as I am able. Furthermore, it is worth noting that encounters are not monolithic in scope, rather "there is a whole scale of possible meetings that ranges from the quite trivial to the extremely significant" (Marcel 1950, 137f). I certainly may encounter Others in a very trivial way, as so many faces in a crowd, or even in an objectified way, as a waiter or server or the like, but I may also encounter an Other as someone significant and important. Marcel provides a beautiful description of just such an encounter that is worth quoting at length:

> All these unknown people present themselves to us, in fact, as mere bodies occupying a certain share of space in the *lebensraum* in which we have to maintain our own share of space and through which we have to thrust our way. But it is enough for some small thing to happen, something which is objectively speaking nothing at all, for us to transcend this subhuman level: for instance something about the tone of voice in which someone in the crush says, "I beg your pardon," or perhaps because of something about the smile accompanying such a simple phrase there is a sudden spurt of clarity, of a clarity that has nothing in common with that of the intellect, but that can somehow light up, as a flash of lightning would, the obscurity . . . through which we are groping our way. (Marcel 1950, 138)

Marcel's phenomenology of this chance encounter is remarkable in itself, but even this need not be the extent of the encounter. It may happen that after such an uncommon experience, I encounter the same Other again, in a different place. If such an event should occur, both of us may be led to "creative development" wherein our relationship grows and the Other becomes for me more than just any other (Marcel 1950, 138). That encountered Other may eventually become for me a close and dear friend, a confidant, or beloved.

Should the Other become for me more than just any other, she and I will develop a sort of intersubjectivity, a concept entirely foreign to Sartre's conflictual subject-object encounters. Because encountering the other occurs to different degrees and with more or less intensity or success, so does intersubjectivity. That is to say,

> There is a hierarchy of choices, or rather of invocations, ranging from the call upon another which is like ringing a bell for a servant to the quite other sort of call which is really like a kind of prayer. . . . In invocations of the first sort . . . the Thou we are invoking is really a He or a She or even an It pragmatically treated as a Thou. (Marcel 1950, 179)

There will be some Others who arrest our attention in such a way that our very way of being-toward-them is of an almost religious dimension, though these will be infrequent but profoundly meaningful, and there will be others whom we see in an almost objectified (though not hate-filled) way, as occupying a role; the person whom I ask directions from on the street, for example, occupies the role of "signpost" or "tour guide" (Marcel 1950, 179). The Other, in this case, does not become an object, but he certainly is not encountered in his depth as "one-like-me."

This suggests that there are various degrees to which we might *encounter* an Other as well as degrees to which we might have *intersubjectivity* with an Other. On the one hand, we may encounter others as morally insignificant fellow subjects: each person *is* a person, and I can recognize this; however, none of them in the crowd is particularly interesting *to me*. On the other hand, I can encounter a particular person in a crowd as unique and one whom I may come to know in a more meaningful way. On the one hand, my interaction with an Other may be merely a formal exchange of information or service. On the other hand, it may be a mystical experience of deep interconnection.

Facial recognition software falls on "the low end" of both the encounter and the intersubjective experience; it prevents the chance encounter of the Other, and it risks alienating ordinary intersubjectivity. A stranger using facial recognition, uninterested in me for anything more than my task as a coffee barista, who calls me by my first name or talks to me about things that deeply interest me can feel like a violation, a forceful appropriation of who I am without first being open to the possibility of real encounter and intersubjectivity. Furthermore, the chance meeting of a brand new Other, the exploratory dimension of new intersubjectivity is lost when I have before me information about the Other's name, close friends and contacts, employment, living arrangements, and habits. The evolution of intersubjectivity is impossible because, as one using facial recognition software, I am presented from the beginning of our encounter with a large amount of information labeled as "comprehensive" of the other person, which hinders the chance of spontaneous growth or discovery.

IV. LEVINAS AND RECOGNIZING THE FACE

A final question of concern, after the issue of the encounter and intersubjectivity, is the question of the basic claim of facial recognition software, that is, does it allow us to recognize faces? To answer this, we must understand what the face is that we might encounter. Sartre and Marcel do not give us a clear account of this, so we turn now to Emmanuel Levinas, for whom the face is crucial.

For Levinas, we do not encounter Others as objects, viz. Sartre, but rather as naked faces. The naked face, stripped of any dressing or pretention, addresses us. "In [the face], the infinite resistance of a being to our power is affirmed precisely in opposition to the will to murder that it defies, because, being completely naked—and the nakedness of the face is not a figure of speech—it means by itself" (Levinas 1998d, 10). The naked face, devoid of any sort of mediation or representation, speaks to us. "Human nakedness calls upon me—it calls upon the *I* that I am—it calls upon me from its weakness, without protection and without defense, from nakedness. But it also calls upon me from a strange authority—imperative, disarmed—the word of God and the verb in the human face" (Levinas 1998c, 198). To encounter the face of the Other, I have to encounter the vulnerable, un-covered, un-protected face. Furthermore, this naked face addresses me, calls out to me in its nakedness and in its Otherness, and as a naked face, it cannot be ignored.

As Levinas illustrates, the naked face of the Other *does* show up to as something-that-can-be-killed, as an opposition or someone to be subjugated as in Sartre, but the demand that the naked face places on us is otherwise—it is the demand of responsibility for the Other.

> Responsibility for the other man . . . would take place in what I call an encounter with the face of the other. From behind the bearing he gives himself . . . in his appearance, he calls to me and orders me from the depths of his defenseless nakedness, his misery, his mortality. It is in the personal relationship, from me to the other, that the ethical "event," charity and mercy, generosity and obedience, lead beyond or rise above being. (Levinas 1998a, 202)

In encountering the face of the Other, I am burdened to answer her call. Her naked, unmediated face addresses me and calls for my response. Thus, my encounter with the Other as face is not one of subject-object, wherein I can only be subject by making the Other object; it is rather one of call-response, where the Other's face calls to me, and I must respond.

Not only does the naked face of the Other call to me, but it is only as face that I am at all able to encounter the Other. The encounter of the face of the Other, as a naked face and as Other, resembles Marcel's deeper notions of intersubjectivity. The encounter of the face of the Other is encountering "language before language," in other words, Scripture (1998d, 199). Thus, the intersubjective experience, the face-to-face encounter with the Other, unmediated and unspoken, is a religious experience, one that resembles prayer. "'Religion' remains the relationship to a being as a being" (1998e, 8). Encountering the *face* of the Other is an encounter with divinity, not an encounter with just another object in the world, nor an encounter with a third-person subject, some "he," "she," or "it." It is not the case that I am merely burdened by the Other: it is only through truly encountering the Other that I encounter the infinite. The Other both compels me and transforms me. Furthermore, this encounter with the Other is possible with any true Other, from the most innocent babe to the most wretched brute (cf. Levinas 1998e, 231). This means that, not only is it possible that I may encounter *some* Other, as Marcel informs us, but that *any* Other, whom I truly encounter as face, becomes for me the experience of the divine.

Levinas is not content to leave the encounter at a quasi-mystical level. The response to the Other places on me an ethical demand. When I encounter the Other, I am given a singularizing ethical duty, "as if, for all eternity, the *I* were the first one called to this responsibility; nontransferable and thus unique, thus *I,* the chosen hostage, the chosen one" (Levinas 1998e, 227). The naked face not only invites my encounter, but demands it, and in so doing, demands my responsibility for the Other. Thus, the "ethical event" occurs and the presence of the Other thrusts upon me the ethical duty I am compelled to follow (Levinas 1998a, 201). According to Levinas, this responsibility impels me to care for the Other, ensuring that the other lives, even at the cost of my own life. Further,

> The priority of the other over the *I*, by which the human being-there is chosen and unique, is precisely the latter's response to the nakedness of the face and its mortality. It is there that the concern for the other's death is realized, and that "dying for him" "dying his death" takes priority over "authentic" death. (Levinas 1998b, 217)

Thus, the naked, unmediated face of the Other—*any* Other—calls to me, encounters me religiously, and demands my responsibility, that is, up to and including dying for the Other.

At this point, we are ready to put forward an idea of what the face is. The face is not a mere collection of eyes, nose, mouth, ears, and other features; the face is our way of truly encountering the Other—the one who is like me, yet wholly Other than I am. The face, that by which we encounter the world and Others encounter us, betrays our vulnerabilities and our weaknesses, yet it also demands our respect and our response. The face appears to us stripped of all pretentions, all attributions, and all clarifications: it is utterly naked. Yet the face calls to us, addresses us, and commands our response—even dying for it. Finally, the face is the place where we encounter the divine, the unspoken word of God placed before us. Each face has the grandest mysteries locked within it, the truth of our being and the infinite, but these mysteries require our attention and our understanding of the Other as the naked Other who calls to us. This is true of any Other whom I truly encounter as Other and not as object or obstacle. The face is a totality; it is vulnerable and divine and it thrusts upon the observer ethical obligation.

The question of facial recognition should now come strongly into relief for us. Facial recognition technology recognizes faces in only a limited sense, and, we might say, does not really recognize *faces* so much as features of a face. For facial recognition software, the face must become an object, something measurable and classifiable. In order to identify "who" a person is, facial recognition must ignore the Other *as Other*. Not only does this occur on a technological level, but it also occurs on a personal level as well: the user is hindered from encountering the Other insofar as the face of the Other is mediated through technology. For Levinas, we encounter the *unmediated*, naked face of the Other, a face that is not presented to us as the ordinary subjective Other, per Marcel; as object, per Sartre; or as some collection of facts and statistics, through facial recognition technology. The face of the Other, through facial recognition, is reduced to an object with attributes: at best, a formed substance. The ability to encounter the Other through facial recognition is circumscribed by software that purports to present the Other to the user, effectively preventing the Other from showing up as Other.

V. CONCLUSION

By reducing the Other to object, facial recognition prevents us from encountering the face (Levinas) of the Other and all that such an encounter entails. At best, it seems I can only encounter the Other in an "indifferent" sense—as a role or a co-subject who can never be a Thou. Through a mediated encounter of the Other, I cannot have a truly religious experience of the Other, as site of the divine, because I cannot *discover* the Other as wholly Other; my interaction may not be "blind," but it will also not be enlightening. Further, the inability to encounter the face of the Other entails the inability to respond to the Other. Levinas claims that meeting the face of the Other is "moral consciousness itself" (Levinas 1998d, 11), but this consciousness is blocked when the face of the Other is obscured through the machination of facial recognition software. Such technology leaves me with perceiving the Other as object, that which "is what it is" and is not a subject like me who can "put her past out of play." *Facial recognition reduces the Other to that which can be manipulated and used*, the being of which is instrumental, and is therefore subjected to the will of a subject like me.

Moral knowledge requires our encounter with the unmediated face. If all faces and all interaction are mediated through facial recognition technology, we will no longer respond morally to Others because we cannot know our moral obligation to them. The nakedness of the face of the Other, which nakedness impels my response, disappears when the Other is objectified through facial recognition, and thus my ethical response evaporates.

Similarly, the Other is unable to respond to me insofar as intersubjectivity requires both persons to be able to recognize the subjectivity of the Other. Thus, all sense of ethical responsibility and intersubjective recognition is made impossible through facial recognition software; all that is left is conflict and objectification, and our moral knowledge of the other is stolen away from us by the objectification and circumscription of the Other.

NOTES

1. I recognize that this claim is itself problematic and warrants its own careful investigation (presumably through the work of Foucault, who brought to our attention the dangers of the technology of surveillance in *Discipline and Punish*), but I leave this claim in its controversial status because the developer of NameTag boasts about this capability.
2. One notes that Sartre is deeply indebted to Hegel's *Phenomenology of the Spirit*, especially sections 431 and 433, wherein Hegel asserts that consciousnesses battle each other to assert themselves in the "master-slave" relationship.

Chapter 17

Situated Mediation and Technological Reflexivity

Smartphones, Extended Memory, and Limits of Cognitive Enhancement

Chris Drain and Richard Charles Strong

In 1928, Paul Valéry speculated that "like water, like gas, like electric current which are brought from far away and in to our homes, responding to our needs through almost zero effort, so we shall be served visual or auditory images which will appear and disappear at the smallest gesture" (Valéry 1960). This passage, partially incorporated into Walter Benjamin's "The Work of Art in the Age of its Technological Reproducibility," illustrates how Valéry's imagination did not go far enough (Benjamin 2006). Today the effortless command of "visual or auditory images" is not limited to passive sensory experience. Indeed, with the ubiquitous computing afforded by the smartphone, it is available for many anytime, anywhere.

In 2014, 58 percent of Americans had smartphones according to the Pew Research Center. Projections compiled by data firm Emarketer show that by 2016, 20 percent of people in the world will own a smartphone. These numbers present a slow and silent revolution in everyday ubiquitous computing. For many, smartphones have become indispensable for work and leisure, for maintaining human relationships, and even for trying to find one's "soulmate." In light of smartphone technology, there has been and continues to be a nontrivial modification of what humans are, how they act, and what they know. No small amount of ink has been spilled prognosticating what the future of humanity will entail with respect to digital technologies (Fuller 2011). However, we insist that even making sense of the present is no small task. The smartphone is not the harbinger of changes for humanity. It is the instrument, evidence, and scaffold of changes already wrought, of practices already modified, of memory and perception already altered.

Central to our project is the foregrounding of the notion of technological reflexivity. Technological reflexivity refers to the way in which technological practices have a bidirectional causal influence and on human agents. Such bidirectionality shouldn't be conceived of only in the manner of traditional propositional logic as an "if-and-only-if" between two sets of terms. In accordance with but also extending beyond the thematic concerns of the paradigms of situated cognition and actor network theory, the "directionality" in bidirectional causation can refer to higher and lower order levels of causal explanation—from institutional levels acting causally on individual actors from the "top-down," to individuals causally acting on society as a whole "from the bottom up," as well as actors affecting each other "horizontally," a process which itself alters the causal feedback of the system as a whole (Sawyer 2005).[1] Looked at from this latter horizontal aspect, technological artifacts occasion the hybridization of human and nonhuman agency, which in turn endows both the user and the device with new ranges of possible action (Malafouris 2013). Put another way, the situated potentials for action between material things in the world and the interactional processes thereby afforded need to be seen as not only constituting the possibility of agency, but thereby also comprising it. *Eo ipso*, agency must be de-fused from any local, "contained" subject and be understood as a situational property in which subjects and objects can both participate. Any technological artifact should thus be understood as a complex of agential capacities that function relative to any number of social and material factors.

Keeping in mind that we are co-constituted by webs of relations involving increasingly complex collections of artefacts, networks, niches, and communities of practice, our investigation will be guided by interrogating the functional potential of a thing that in the last fifteen years has seamlessly worked its way into the everyday life of millions of human agents. This "thing," the smartphone, is merely a nodal point in a highly complex network. Recognizing this massive "collective," we nevertheless want to show some of the ways in which something as seemingly mundane as a smartphone can reflexively alter the range of actions available to a cognitive agent (Latour 1994). Specifically, we hope to shed light on some of the social-cognitive consequences of technological mediation by looking at the complementary, if not mutually implied, domains of memory and knowledge.

I. MEMORY AND SOCIAL-ARTIFACTUAL MEDIATION

In this section, we will look at some of the impacts that smartphone use has on memory. While the case has been made that ubiquitous computing has already altered the cognitive functions of attention and short-term memory, we will be limiting our discussion to the effects of smartphone use on explicit long-term memory (Hayles 2007; Carr 2010). The relevance of smartphones with respect to semantic memory will be touched on here, but more directly addressed in the next section insofar as it relates to our everyday knowledge and beliefs. We choose episodic memory because its autobiographical, apperceptive, and temporalizing aspects relate most clearly to our larger thesis that the smartphone is a potent material agent that significantly alters the range of cognitive actions available to its user (Campbell 1997). And with respect to memory, such reflexive technology plays a vital role in the social construction of the self and collective remembering.

I.1 Situating Memory

From the perspective of situated cognition, memory is viewed not simply as a retrieval system but rather as a reconstructive process that depends on ever-changing contextual inputs. Literally *extending* connectionist and constructivist schemas that explain remembering as a process entailing the activation of superimposed neural "traces" whose distribution and potency depends on the history of the network, a *situated approach* adds that extracranial resources and processes can be just as integral to the (re)construction of memory (Sutton 2009, 219). The move from a connectionist-constructivist to a situated account does not deny that interior neuro-cognitive processes can be sufficient for acts of memory, but rather seeks to extend the model. On an internal scale, remembering is a temporally situated and contingent process of reconstructing distributed and sedimented neural traces. On an external scale, remembering is a materially extended and socially distributed process that relies on bodily and nonbodily scaffolds, artifactual assemblages, exogrammatic symbols, signs, and icons, as well as joint-attentive processes of social and dialogic cueing (Vygotsky 1997; Clark 2010b; Donald 1991; Aydin 2015; Scalambrino 2012; Hirst and Echterhoff 2012).

The functional scope of the smartphone is vast, expanding well beyond its original telephonic intention. A quick look at one's local bus stop, café, or even Tahir Square will illustrate quickly that most people use their phones for a variety of tasks: People converse through text, take and share photos, record video, update social media, etc. A common aspect of these actions is that they all include a recording or documenting element. We record with smartphones and communicate those recordings. The phone can thus be looked at with respect to memory as a prosthetic corrective to the deficiencies of bare intracranial memory, one that literally extends our cognitive capacities.[2] It would be a mistake, though, to view such a corrective as issuing from its functional durability. It is possible that hard plastic, liquid crystal, and microchips outmatch the durability of neuronal gray matter. But as anyone who has lost their device to an accidental drop can attest, a smartphone can be rendered dysfunctional or un-functioning in an instant.

With the advent of cloud storage, however, such dysfunctionality is only a temporary misfortune. Such information is never really lost so long as one makes her monthly payments.

I. 2 Memory and Reflexive Mediation

The smartphone functions as a reflexively mediating device in the following senses:

1. It *becomes incorporated* within the assemblage of bodily appendages, environmental features, and artifacts that we encounter in everyday life, to the point where the phone can be considered as a prosthetic extension of ourselves.[3]
2. It *incorporates us* within a socially mediated system of symbols, ideas, and information broadly construed.

The ramifications of such cognitive coupling for how we remember can be looked at from three levels: first, with respect to the individual's own relation to her memories; second, with respect to social forces that bear on such memory making; and third, with respect to how we collectively remember. It is important to keep in mind that while these levels are heuristically isolatable, they function in tandem, each reciprocally affecting the other.

The smartphone directly affects how we construct our own personal narratives and self-identities in that it allows for the constant volunteering of information (e.g., texted thoughts, images), information which may be accessed at another time in almost instant fashion without the risk of organic or natural deterioration. In other words, the smartphone virtually radicalizes the causal directionality of externalized memory. In terms of the causal direction of agent-to-memory, the smartphone makes two great strides. One, not only does it make available a potentially unlimited (or at least a potentially unimaginable) amount of "storage" in which externalized information may be deposited, but because of its digital reproducibility such storage safeguards this information from natural "forgetting" in a way unrealizable before the advent of cloud technology. Two, the portability of, and in most cases, the perpetual online connectivity of, the smartphone render the pragmatics of such memetic exteriorization transparent as they do ubiquitous. This works just as well from the top-down. With respect to the causal direction of memory-to-agent, the smartphone enhances the recall potential of such stored information for the same reasons. How one extends her memory and how that memory is reconstructed is thus deeply altered with smartphone use.

The smartphone's function of person-to-person communication also can alter memory construction. The fact that memory creation can depend on social and joint-attentive processes is a touchstone of the situated paradigm (Sutton et al. 2010). Broadly, the idea is that dialogic cueing not only aides memory recall but can actually comprise it. For example, a husband and wife can reconstruct past events through shared conversation—neither can independently recall a past event but through shared conversation they are able to do so (Hirst and Echterhoff 2008). In such a case the husband and wife exist as a joint agent with respect to memory construction. Adding a smartphone to the mix only enhances this agential capacity. Surely adding smartphones to a standard conversational routine enhances the recall capacity of the joint-agent in question. But if we consider that many relationships are first created and then maintained through smartphone use (e.g., through access to social media), then the possibilities for memory construction are expanded exponentially. Instead of human-human agential hybrids we have human-smartphone-human assemblages, with each constituting element reciprocally affecting the other in memory construction and recall. It is at this level of analysis where interpersonal and joint-attentive approaches to memory blur into the general social construction of memory.

There is considerable philosophical debate as to how an epistemic or cognitive status can be predicated of a group or collective, and a proper treatment of the nuances of collective memory is unfortunately beyond the scope of this chapter (List and Pettit 2011; Tuomela 2005; Searle 1995). But whether a plural subject of memory is thought of as the aggregate sum of individuals or as an emergent entity in its own right, we take it as uncontroversial that smartphone technology shows to be a vital component of how and what we collectively remember (Theiner and O'Connor 2010). Given that any event can at any time be recorded and shared by and for a multitude of agents, what and when we remember has drastically changed. For example, the Ferguson protests, the events of the Arab Spring, and the tragic events of 9/11 were documented and shared in real time by a variety of sources, each contributing to our shared histories, images, and narratives. Whereas traditional channels of journalism acted as gatekeepers, framers, and fact-checkers of such information, such editorial mediation is bypassed through the tandem use of smartphones and of social media sites such as Youtube, Facebook, and Reddit. Indeed with the smartphone in hand, a "go-to" protest chant has become, not infelicitously, "the whole world is watching." This is not much of an exaggeration given that any recording can become publicly available in a matter of seconds.[4]

II. "SMART"PHONES AND EVERYDAY MODES OF KNOWING

In keeping with our proposed framework, we'll consider what and how an agent coupled with a smartphone can know and indeed also how such knowledge may in some cases be enhanced, diminished, and otherwise altered. The scope of our analysis will be limited to a few of the more common ways in which everyday human knowledge and smartphones are reflexively related.

II.1 Enhancements in Everyday Knowledge

We routinely use smartphones for acts of identification, navigation, reference, corroboration, and computation. Given that a smartphone can access knowledge with fine granularity in mere seconds wherever one happens to be, such an enhancement marks a radical turning point in the range of such aforementioned "acts of knowledge." The sheer speed of access and the amount of facts, maps, figures, statistics, and probabilities doesn't simply produce a quantitative shift in our potential for knowledge—enough quantitative alteration in this case constitutes a qualitative change in who, what, and how we know, as well as where we can skillfully go. The everyday character and ubiquity of the smartphone tends to cover over the radicality of this technological alteration to our cognitive and agential potential.

The human-smartphone hybrid diverges greatly from merely biological ways of knowing and encountering spaces and places. Take the example of smartphone map applications.[5] Limiting our analysis to the user-end experience, one can see that these maps obviate the need to acquire expert spatial-cultural command of a place. We know that expert knowledge of a city correlates to physical changes in the hippocampal gray matter (Maguire et al. 2006). With map applications, however, such biologically grounded skills are largely transferred to the smartphone. This lightening of the cognitive load brings with it both advantages and disadvantages.

The advantages are obvious to anyone who's used a smartphone to get around in a new city: a complex and novel environment becomes easily navigable with recourse to a map application. Such applications allow one to switch between multiple layers of navigational information from different and sometimes competing sources, improving drastically upon precellular modes of skillful navigation in unfamiliar places. If you take away the phone, then overall navigational competency drops significantly. However, such enhancement may have a deleterious effect on the experience of space and place in that any robust sense of specificity is dissolved and leveled over.[6] Despite the ability to toggle between a nominally diverse offering of information, such maps are largely the same for all users in that they tend to circumvent the social and individual thickness of meanings afforded to an embodied, phoneless, social agent. Smartphone users are also less likely to converse with or ask directions from those around them. This makes possibly serendipitous or friendly face-to-face social interactions less likely, a prime example of what Sherry Turkle calls being "alone together" (Turkle 2012).

II.2 Unintended Consequences for Learning and Knowing

Smartphone use brings with it unintended and inconspicuous challenges to how and what one knows. In college classrooms, for example, many students constantly use their smartphones for personal communication and entertainment in a way that seriously disrupts their ability to learn and to retain lessons (Smith 2011). This is perhaps in part due to generational changes in synaptogenesis that favor multitasking and diminish the ability to focus one's attention in a sustained manner (Hayles 2007). Lessons often aim at inculcating dispositional intellectual skills. This is not the sort of *processual* knowledge that can easily be accessed and assimilated via a smartphone. Nobody can tweet their way to advanced literacy, ask Siri how to think critically, or use a smartphone to grant complex, skill-dependent thinking, speaking, and interacting. In turn, this may chip away at some of the intellectual skills and *savoir-faire* that are necessary conditions for human flourishing in a democracy (Plato 1997c; Dewey 1997; Nussbaum 1997). Though these trends may be in step with other changes in our built environment, which encourage distraction and discourage deep attention, this is unfortunately one way in which the epistemologically "enhancing" capacity of smartphones undermines the gains that it grants (Strong 2014). However, it is important to note that the mediating role of the smartphone in educational settings is not as one-dimensional as it may seem. It is possible to imagine a scenario, for instance, in which a student learns more by watching Michael Sandel's famous lectures on justice from a smartphone than a student who was there in person but was distracted by her own smartphone.

Being aware of and not overstepping one's epistemological limits has been a perennial theme in philosophy since at least Socrates, and such a problem is not lost on the smartphone user (Plato 1997a). That is, recent experimental findings in psychology suggest "that searching the Internet for explanatory knowledge creates an illusion whereby people mistake access to information for their own personal understanding of the information" (Fisher et al. 2015). Thus using smartphones to access information contributes to the overestimation of "internal" knowledge and understanding of information. This case is interesting in that even in its absence the smartphone can have an effect on its user. Even when a phone is not at hand, the mere fact of being accustomed to its use renders an agent overconfident in estimating her own epistemological limits (Ward 2013).

III. CONCLUSION

The ubiquitous computing engendered by smartphone use has indeed already wrought major changes to how we act as social, cognitive agents. In the first part of this chapter, we argued that in keeping with the thematic of situated cognition, the smartphone allows for memory to be extended and distributed in a manner hitherto unimaginable. The content and means of identity creation becomes radicalized when considered from the perspective of a hybrid artifactual agent. Such a shift consists in the reciprocal nature of memorialization insofar as such a process is grounded on the constant access to and creation of documents, whether imagistic or textual in nature. With a smartphone an agent's personal memory is augmented and enhanced, and insofar as that agent is merely one part of an assemblage of likewise enhanced agents, interpersonal and collective modes of memory are similarly altered in their potential for self and social identity construction. In the second part, we addressed how, when considered epistemologically, the "enhancements" and modifications that follow from such hybrid agents are uneven and problematic. We showed that although perpetual access to information alters what and how the average human agent knows, it does so *not* in a purely progressive or enlightened way. The smartphone alters those who use it by diminishing their attention in formal learning settings and by producing agents who tend to overestimate their own "internal" knowledge. In sum, we hope that this modest contribution adds to the larger picture of how humanity is not a pre-given notion and that alterations to seemingly fixed human capacities such as knowing and remembering are intimately bound up with and altered by technological practices.

NOTES

1. We realize the analytical dangers involved in grouping such diverse theoretical paradigms. But we hope to appeal to a certain commonality between the situated paradigm in cognitive science, emergentism in the philosophy of social science, and Latour's actor network theory (ANT). All three paradigms avoid traditional binaries setting inner against outer and subjective against objective. Sawyer's theory of social emergence adds to ANT and situated theories of cognition the idea that actions, whether social or cognitive, are subject to constraints and/or causal forces at a variety of ontological levels each with its own relative state of stability and each in a relation of reciprocal causality and/or constraint with the rest (Sawyer 2005, 189–220). Such a theoretical blending, we hope, enriches rather than confuses our discussion.

2. We are not advocating that this is always a good thing. From Nietzsche to Charlie Brooker, many have pointed out the negative consequences of *not* being able to forget (Nietzsche 1980; *Black Mirror* 2011). Contemporary legal arguments over the "right to be forgotten" illustrate that this is ever more a concern in the age of digital media (Rosen 2012).

3. The term "personal electronic device" reflects not just ownership but the deeper attitude that a phone is literally a part of one's person. Many panic at the thought of

leaving theirs behind and "phantom vibrations" have become a bodily experience unique to the twenty-first century (Drouin et al. 2012).

4. The effects of such rapidly ubiquitous information are not always positive with respect to narrative construction. Take the case of the Reddit imbroglio of 2013. In their zeal to aid in the capture of the perpetrators of the Boston marathon bombing, users of the social media site collaborated with others users' smartphone footage from the scene to mistakenly identify several suspects. Such crowdsourced sleuthing ended up stalling the legitimate channels of justice (Kaufman 2013).

5. We ought to point out that the smartphone's relation to knowledge is not merely one of access but also of production (Brabham 2013). However, looking at the use of smartphones to produce maps and/or informatively tag such maps through crowd-sourcing is beyond the scope of this essay.

6. These maps are often owned by for-profit companies and display advertising nudges designed to make the user purchase goods and services. These maps can even track one's movement for the sake of delivering individually tailored advertisements or other behavioral nudges. While beyond the scope of this chapter, we recognize that this is a hugely problematic issue of central importance to a comprehensive account of everyday technological practice.

Chapter 18

The Vanishing Subject

Becoming Who You Cybernetically Are

Frank Scalambrino

"To impersonate is to play at being another; to personate is to play at being oneself" (Browning 1990, 172). The first part of this volume includes the examination of the normative social dimensions effected by the technological mediation of knowledge, and the second part includes the examination of the changing conceptions of humans and humanity influenced by technological mediation. Heuristically speaking, then, we may say that the first part of the book places emphasis on the relation between technological mediation and epistemology, understood as first and foremost *social epistemology*, and the second part of the book places emphasis on the social effects of technological mediation understood in terms of a relational ontology and applied cybernetics. Whereas the first part may be characterized as an examination of the social effects of a technologically mediated influence to human regulative ideas, the second part may be characterized as an examination of the existentially, that is, socially and individually, constitutive effects of these human regulative ideas altered through the influence of technological mediation.[1]

The terminology of "regulative idea" comes from Immanuel Kant (cf. 1998), especially his 1781 *Critique of Pure Reason*. Simply put, a regulative idea is an idea used to regulate a human relation, whether it is to a thing, a person, or a situation. How these ideas may become constitutive of human lived experience is quite straightforward. For instance, if the idea regulating a person's relation to food is such that the person thinks that consuming food will result in undesired consequences, then the person's resistance to eating both is influenced by that regulative idea and will have existential consequences (cf. Kroon and Perez 2013). Such existential consequences not only eclipse the present moment for a person, they also extend into the future. Consider how anorexia functions like malnutrition both physically and psychologically.

Moreover, it is important to note how regulative ideas become constitutive of "lived experience." For it is also the case that, because the idea regulating one's relation determines meaning by also regulating the meaning-making process involved in the experience (of reality), the person lives "as if" the meaning determined by the active regulative idea were true of reality (cf. Vaihinger 1949). In regard to changing conceptions of humans and humanity, then, this chapter goes toward a discussion of how the technological mediation of (social) reality influences personal regulative ideas, which as the individual lives into and through them, participate in the constitution of the person's lived experience of (social) reality. If we focus just on the aspects of lived experience contributing to one's self-awareness, then we notice a constitutive reinforcement at work, so to speak, through the technologically mediated regulative idea, constituting one's personal identity (cf. Kant 1960, esp. 22–23).

Though it is possible to think of a regulative idea as part of a larger thought constellation, two points of clarification need to be made. First, we should not invoke notions of an "Unconscious" (cf. psychoanalytic "Other") regarding the nonactive constellation for, at least, three reasons: (1) the notion of an Unconscious would itself be a regulative idea, (2) whether understood biologically or socially, psychoanalysis takes the Unconscious to be deterministic, and we are, rather, interested in affirming human existential freedom; and (3) though it is possible to think of a larger nonactive thought constellation encapsulating a regulative idea, such a thought constellation would be momentary. That is to say, any such thought constellation, because it pertains to *lived* experience, should be understood as temporally contained. Therefore, were such a thought constellation to be present across time, something would need to sustain it, that is, to invert the existentially originary particular to general temporal expression, which is precisely why psychoanalysis is by its nature a high-frequency, long-term "therapy" (cf. Ellis 2002; cf. Torrey 1992; cf. Roustang 1982 and 1990). Second, thinking of a nonactive constellation may push us into a cybernetic understanding of the thought constellation, for example, as "the Unconscious," which as will be discussed (and insofar as it is a theme of this volume), is precisely what social epistemologists and philosophical theorists have indicated is one of—if not the one—major problems with technological mediation.

FROM THOUGHT MEMES TO THE ME GENERATION: EVERYONE'S A TECHNOLOGICALLY MEDIATED STAR

There is no such thing as a "gap year." "I know, right?" "Just saying." These locutions are examples of "thought memes" (cf. Dawkins 1989, 192; cf. Deleuze 1986, 205). "Meme" refers to the transmission of an information pattern across space and time that is culturally or socially based. The idea is that despite content and context changes, expressions are being imitated and repeated. Across the repetitions, existentially speaking, there are lived experiences at stake. On the one hand, the perhaps simplest way to explicate such a claim would be in terms of "habit." On the other hand, it is important to notice how such repetitions can occlude or conceal one's own-most potentialities. These potentialities are existential, and beyond a mere opportunity cost-benefit analysis, when technological mediation has influenced a circumspective shift of individual being toward the actualization of virtual reality, the subject vanishes amid a virtually limitless horizon.

Two points of clarification are needed: (1) a brief discussion of the difference between potentiality and virtuality and (2) the relationship between habit, thought memes, and existential freedom. First, consider how we sometimes say that the figure is already in the marble or the statue is already in the block (cf. Deleuze 2002, 82). The idea being that so long as one has the ability to sculpt or carve the desired end result, that is, accomplish the material instantiation of one's regulative idea, then the end is virtually already present. When we consider an actual moment of existence, we may ask if the agents involved in that moment have the *potential* to accomplish some activity. In regard to virtuality, we still consider an actual moment; however, we envision the possibilities which may actualize. Deleuze refers to the virtual as "a pure multiplicity in the Idea" (Deleuze 1994, 211). On the one hand, we have a regulative idea in relation to the marble or block, and the determination of "the virtual *content* [emphasis added] of an Idea" is the idea's differentiation (Deleuze 1994, 207). On the other hand, "the actualization of that virtuality into species and distinguished parts" may be thought of as the physical instantiation of the idea in reality (Deleuze 1994, 207).

Second, recall from Aristotle's *Nicomachean Ethics* the three criteria for virtuous action: (1) freedom of the will; (2) knowledge of the situation; and (3) the tendency to produce the action. In other words, for Aristotle, one is not supposed to be able to be a virtuoso by accident or despite performative ignorance (Aristotle 2009, 1105a30–35). Interestingly, then, habit pertains to the tendency criterion and thought memes pertain to the knowledge criterion. Though it is beyond the scope of this chapter to explore, Honoré de Balzac's 1831 novel *The Magic Skin* may be seen as a reading of technological mediation (i.e., the magic skin) through Aristotle's criteria; however, whereas the knowledge criterion for Balzac was understood in the context of philosophy, the contemporary context would be cybernetics. On the one hand, what this means is that insofar as one's knowledge-base is constituted by thought memes, then one's freedom may be seen to be similarly infiltrated. When confronted with a situation, such a person's (regulative) idea of the situation is virtually not his own. On the other hand, the deeper existential problem may be seen at the intersection of these two points of clarification.

Just as the labeling with which one understands existence may be meme based, so too one's view of the horizon of possible actions may be based in regulative ideas differentiated in terms of technological mediation. Insofar as this is the case, then one may be oriented toward actualizing the virtual dimension, rather than the dimension of potentiality. Beyond the alienation and inauthenticity involved in regulating one's life in terms of what may be *virtually possible*, given technological mediation over what one has the power to actualize, the former relation to existence may be populated with illusory ends. Whereas power expression involves the actualization of some content, retreating à la reduction to some genre of gratification reveals the virtual end to be parasitic more than it justifies its presence as a regulative idea, that is, the mere expression of power is not to be equated with existential freedom (cf. chapter 9 of this volume). In other words, once cybernetic or meme-based knowledge permeates existence to the point of habit, then two of the three criteria have been infiltrated by technological mediation such that it is as if the will is redirected toward a virtual dimension. It is not that one cannot express the content differentiated through technological mediation, rather one's point of departure in determining existence from moment to moment has shifted.

A more concrete example of the above discussion of the vanishing subject may be seen in Ernst Jünger's discussion of the relation between the globalization resulting from applied cybernetics and the self-understanding and public-based self-awareness of individuals as "workers," that is, human resources, standing-reserve, or part of a stockpile. According to Jünger,

> Wherever man falls under the spell of technology, he finds himself placed before an unavoidable either/or. This means that either he accepts the particular means of technology and speaks their language, or he perishes. If one accepts (and this is an important point), one makes oneself not only the subject of technical processes, but simultaneously their object as well. (Jünger 1963/1983, 273; cf. Scalambrino 2015a)

Notice how Jünger's reference to speaking the language of technology may be seen as a gesture toward an orientation regarding the instrumentally best, that is, the most efficient, cybernetic thought memes. Through the work of Deleuze and Guattari we might articulate Jünger's reference as a "power take over by a dominant language within a political multiplicity" (2005, 7). Reminiscent of the "steersman" and cybernetic discussion from chapter 9, notice Jünger's discussion of the manner in which the subject becomes object, that is, vanishes.

Being permeated by the language of technological mediation, that is, theoretical cybernetics, involves permeation of the criteria conditioning both the application of cybernetics and the subject's vanishing. Cybernetics is clearly a meta-language; though from the cybernetic perspective regarding language, there would be no meta-language. The term "ambivalent" may be ambiguous in this context, since it is not clear that the subject's ambivalence occurs on the same "plane of knowledge," so to speak. That is to say, the person may be ambivalent between being an authentic person (personating oneself) and being subject to the process of (cybernetic) mediation, that is, impersonating another in the effort to impersonate oneself. Given this characterization, "authenticity" may be said to refer to a resolute resistance to an illusory mediation in the process of being (oneself). Insofar as the effort to impersonate another is the expression of power for the sake of personating oneself, then, just as Jünger characterized it, one is simultaneously subject(ed) to cybernetics and vanishes as subject. Hence, it is as if looking through the all-seeing eye of cybernetics we become part of what it sees, that is, as if we see the steersman seeing us; this self-awareness both determines the actual, virtually limitless horizon of action, and conceals the more ontologically primordial relation to being, thereby concealing one's relation to oneself, that is, one's power center, as the actual point of departure for action.

PRO-TECHNO-CREATION: STEPFORD CHILDREN OF A BRAVE NEW SOCIETY (?)

As "French Existentialist" Gabriel Marcel explained what he called "the Technological Age," "It finally boils down to the question of what modern man, *controlled* [emphasis added] as he is more and more by a technology which he himself has invented, is making of himself" (Marcel 1962, 27). Marcel indicates how "due to the progress of science and of technology" humans are tending toward the preference for the fruits of human mastery of nature, rather than the fruits of nature. In other words, Marcel sees an increasing emphasis on production under human control. And despite noting "the act of procreation can in no way be compared to a manufacturing process" (Marcel 1962, 31), he concludes that "people today will grant themselves the right to manipulate it" (35). Marcel's concluding comment draws our attention to the fact that without some philosophically reflective account of our decisions to perform actions mediated by cybernetics, the default position is that of the naturalistic fallacy. The naturalistic fallacy holds regarding action that "is" implies "ought," that is, the fallacy erroneously suggests because "we can" means "we should." Marcel refers to psychoanalysis as the "tributary of a naturalistic philosophy," that is, essentially the entire psychoanalytic edifice is erected on the naturalistic fallacy, which in its "de-humanizing" of man resonates with, and may even be used to defend, the de-personalizing and de-humanizing (social) effects of technological mediation. Hence, Karl Kraus' à propos quote: "Psychoanalysis is that mental illness of which it believes itself to be the cure."

Emphasizing "chemical control" by pharmaceuticals (cf. Szasz 2003), especially contraception, that is, the birth control pill, Abigail Bray notes, "we desire our oppression because we have come to believe that the practice of self-*control* [emphasis added] is also the achievement of an emancipated autonomy" (Bray 2009, 83). Moreover, this self-control is in the service of a cybernetic knowledge-base commercially looped through technological mediation such that options are available for the subject to cybernetically calculate in terms of a cost-benefit analysis. Faced with choices regarding procreation, it is the very choices that eclipse the nontechnologically mediated relation to the procreative act. In considering the choices made available through technology, whether it be in terms of contraception or the "direct-to-consumer genomics" otherwise known as "designer babies," the cybernetic understanding of the action is intimately involved in what amounts to the applied cybernetics of procreation (cf. Whetstine 2015; cf. Sterckx et al. 2013; cf. Gordon 1999; cf. Green 2007). According to Marcel,

> Life is considered less and less as a benefit. Men of today are much more inclined to emphasize whatever is absurd and hopeless in life, and from this follows the consequence that parents think of themselves as having without justification brought into the world a being who never asked to live, who did not ask to participate in this incomprehensible and all too often sinister game in which they themselves are engaged. (Marcel 1962, 34)

On the one hand, perhaps we may (sincerely) be able to one day, through cybernetics as eugenics, design the trait of hopelessness out of the gene pool, as evidently the "warrior gene" has already been discovered (Tilhonen et al. 2015; cf. Jünger 1963/1983, 287). On the other hand, notice how the options available, given control of the process through technological mediation, have changed the human relation to the process. As we continue to cybernetically mediate our relation to carnal "goals," we may characterize the spectrum along which humans will oscillate from "the drudgery of pleasure" (cf. Guillebaud 1999, 87) to Stepford children. Moreover, with the capacity to eventually reduce or eliminate human psychological and behavioral tendencies, some reflective (public) philosophy is going to need to take place regarding the ethics of such applied cybernetics.

In an Aristotelian key, notice how the efficient cause of Stepford children would be the "ego(s)" of the parent(s). That is to say, parents would be attempting to design the potentialities, that is, the potencies, of their children based in their virtual cybernetic worldview. In terms of the formal cause, the parents would be designing the souls of their children. Framed along the lines of Marcel's critique, such parents would be acting as if they could be the sole authors of their children's souls. Further, because the operative "interests" of parents would be historically, economically, and culturally based, Heidegger might describe the final cause of such Stepford children as being determined by inauthentic market-based (merchant) mentality. That is, assuming, of course, that despite being able to "afford it," parents would not be allowed to genetically engineer serial killer children. Lastly, in terms of the material cause, insofar as the cybernetic worldview is embraced, the material cause might still be an open question (cf. "Soldier 2.0").

ECLIPSE: THE DARK SIDE

The following is an explication of two issues from out of a more survey-like treatment, among others, from elsewhere (cf. Scalambrino 2014d). One more example may help shed light on what is meant by one's "becoming cybernetically who one is." The idea here is to recognize how actual lived experiences are had in relation to artificial, that is, manufactured, and illusory objects of desire. Studies continue to grow showing the use of social media to be associated with personal, societal, and family issues such as, for example, infidelity (cf. Hertlein and Piercy 2006; cf. Cravens, Leckie, and Whiting 2013; cf. Whitty 2005; cf. Young et al. 2000), divorce (cf. Amato and Previti 2003; cf. Tulane et al. 2011; cf. Utz and Beukeboom 2011; cf. Clayton et al. 2013), stalking and surveillance (cf. Hesper and Whitty 2010; cf. Donath and Boyd 2004; cf. Spitzberg and Hoobler 2002), addiction (cf. Clayton et al. 2013; cf. Valenzuela et al. 2009; cf. Chaulk and Jones 2011), body image (cf. Want et al. 2009; cf. Gutiérrez-Maldonado et al. 2010; cf. Fardouly et al. 2015), and depression (cf. Amichai-Hamburger 2011; cf. Jelenchick et al. 2013; cf. Primack et al. 2009; cf. Steers et al. 2014).

From the perspective of a philosophical meta-analysis of the above studies, there seems to be two primary mechanisms at work regarding the technological mediation of interpersonal relations, that is, the production of illusory objects of desire and the narrowing of self-identity required to meet the presentational constraints of the (social) medium. The two are closely related. Though the latter mechanism pertains directly to the artificiality of the mediated relation, presentational constraints of the medium may also account for the "highlight reel" description in the literature regarding the social media sites of others, and as such, the latter mechanism contributes to the mechanism of the former, that is, the production of illusory objects of desire. Moreover, it is perhaps the combination of the two that contributes to an artificial type of competition with others.

How, then, are the technologically mediated social effects thought to eclipse lived experience as one becomes who one seems to cybernetically be? First, insofar as illusory objects of desire are illusory, then the locus of motivation may be attached to the medium. When contrarians speak up in defense of "sacred" social media, attempting to defend social media by claiming such objects of desire are "subjective," we say, "That's exactly what they are." Certainly persons may decide to compete regarding which of their photos get more button pushes, and in the context of the capitalism of the future (cf. Hardt and Negri 2001) we understand there may be money at stake. However, it is also the case that these desires are technologically mediated, manufactured, and despite being fused with natural (think naturalistic fallacy here) psychological dispositions, the pursuit of these desires directs one into a cybernetic dimension, the negative (social) effects of which, since they stem from a virtual dimension, cannot be uprooted.

Second, consider the idea of "image management." Supposing others to be paying attention to your technologically mediated cybernetic self-presentations and retaining the information, then, a social expectation has been created, that is, you are expected to actualize your potential to be who you virtually are. How might such a seemingly simple task be problematic, if not impossible? On the one hand, insofar as the construction of the virtual version of you, that is, who you cybernetically are, depends on narrowing aspects of lived experience so that you may conform with the presentational constraints of the medium, it may be said that your actual lived presence has an excessiveness to it that cannot be reduced away or limited other than in a false cybernetic light. The example that comes to mind involves social media such as those that allow users to come together for otherwise anonymous acts of "cuddling," etc. Despite high definition pictures, one would hope that stench and other aspects of hygiene would be a concern. Yet, such aspects of lived experience are excluded from the technologically mediated relation in which one is making existential choices. On the other hand, as noted above, once you enter into the feedback loop with your own image, that is, the "version of yourself" others access through technological mediation, you are then virtually not yourself, as you strive to impersonate a construction of yourself multiply conditioned by technological mediation.

Lastly, just as the scholarly literature has taken to using the description of "highlight reel" to refer to the contrived nature of self-presentations through social media, so too may the proliferation of an artificial competition be discerned in the very fabric, or "inter(-)face" features, of social media. It is as if the "face of the Other," to turn a phrase from Emmanuel Levinas, in the context of social media refers to the face of the cybernetic steersman. That is to say, *habit*uating to the various actions required to perform functions on and through various social media is tantamount to extending one's performative actions onto a virtual field the rule-structure of which enables the construction of "highlight reels." Just as it is the nature of a highlight reel to contain highlights, competition is already woven into the construct. As if changing our relation to the future, each reel is to better than the last. Hence, habituation to the interface features betrays the seamlessness of being permeated by applied cybernetics.

In this way, it is as if (artificial) competition saturates the relation constituted by the technological mediation of social media. It is well established today that social media, like a cybernetically enhanced version of the elementary school "game" called "show and tell," functions to fuel competitiveness, even among "friends." Yet, what is more, combining the competition relations, that is, the relation with technologically mediated self-presentations of others and the relation with technologically mediated past self-presentations, notice how one's horizon of existential possibilities may be eclipsed. We need not even speak in terms one's "ideal self" as dependent on such technological mediation and virtual reality. At an even deeper level of permeation, regarding how choices emerge for an individual to expend time, one may be seen as placed in the moment-to-moment service of becoming who one cybernetically is. For example, social media-induced depression and jealousy indicate the mobilization of human forces, as such, tantamount to applied cybernetics. We cannot accept the contrarian quibble that no one takes such technological mediation seriously or that it is always already understood to be a falsification of the truth, as if merely for entertainment purposes, since the social effects of the technological mediation are clearly real.

CONCLUSION

The unfreedom, then, associated with applied cybernetics can permeate to the level of constitutive changes for humans in multiple ways. Just as the figure is virtually present in the uncarved block and actually carving it constitutes its physical instantiation, in one's attempt to become who one cybernetically is, one alters one's relation to (social) reality in a way that is existentially constitutive. However, because existential freedom pertains to one's potentiality, the result of cybernetically mediating one's relation to one's existential freedom amounts to embracing a version of one's self determined by the technological cybernetic mediation, rather than one's own-most potential. Such a relational alteration does have constitutive effects. Yet, such effects take hold through the instantiation of mediated habits and mediated knowledge, thereby ultimately redirecting the freedom of one's will toward a virtual reality.

The technological mediation of procreation serves as an excellent example. Questions regarding how to "design" children fall within the context of social epistemology insofar as the answers to such questions are clearly rooted in historical, cultural, and economical motivations. Further, the lived experience of encountering a technologically mediated horizon of existential possibilities shows how such "options" may eclipse what would otherwise be the more originary relation to the procreative action. The structure and mechanism of such eclipsing extends across other human relations dominated by cybernetic technological mediation. This chapter took the effects of a cybernetic technological mediation of one's relation to one's existential freedom as a prime example, suggesting that many other inauthentic activities and involvements hinge on one's cybernetically eclipsed relation to their own existence. In the attempt to become who one cybernetically is, one subjects oneself to cybernetic objectification such that this subjectivity vanishes into virtual knowledge and a habit structure constituted through cybernetically mediated regulative ideas, that is, impersonating another in the effort to impersonate oneself.

NOTE

1. I would like to thank Hannah Lander for her comments on an earlier draft of this chapter.

Bibliography

Achterhuis, Hans. 2001. *American Philosophy of Technology: The Empirical Turn.* Bloomington, IN: Indiana University Press.

Adorno, Theodor W. 1973. *Negative Dialectics.* Translated by E. B. Ashton. London: Routledge.

Agamben, Giorgio. 1999. *Potentialities: Collected Essays in Philosophy.* Translated by Daniel Heller-Roazen. Stanford: Stanford University Press.

Agassi, Joseph. 1989. "The Role of the Philosopher Among the Scientists: Nuisance or Necessity?" *Social Epistemology: A Journal of Knowledge, Culture and Policy* 3(4): 297–309.

Allen, Amy. 2008. *The Politics of Our Selves: Power, Autonomy, and Gender in Contemporary Critical Theory.* New York: Columbia University Press.

Althusser, Louis. 1971/2014. *On the Reproduction of Capitalism: Ideology and Ideological State Apparatuses.* Translated by G. M. Goshgarian. London: Verso.

Altman, Irwin. 1975. *The Environment and Social Behavior: Privacy, Personal Space, Territory, Crowding.* Pacific Grove, CA: Brooks/Cole Publishing.

Amato, P. R., and D. Previti. 2003. People's reasons for divorcing: Gender, social class, the life course, and adjustment. *Journal of Family Issues* 24: 602–26.

Amichai-Hamburger, Yair. 2011. "Depression through Technology." *The New Scientist* 204(2739): 28–29.

Anderson, Kristin J. 2014. *Modern Misogyny: Anti-Feminism in a Post-Feminist Era.* Oxford: Oxford University Press.

Andsager, Julie L. 2014. "Research Directions in Social Media and Body Image." *Sex Roles* 71(11–12): 407–13.

Angwin, Julia. 2010. "The Web's New Gold Mine: Your Secrets." *Wall Street Journal* (July 30): http://www.wsj.com/articles/SB10001424052748703940904575395073512989404.

Aristotle. 1984. "Eudemian Ethics." Translated by J. Solomon. In *The Complete Works of Aristotle: The Revised Oxford Translation.* Vol. 2, edited by J. Barnes. 1922–81. Princeton, NJ: Princeton University Press.

_____. 1995. "Politics." Translated by Benjamin Jowett. In *The Complete Works of Aristotle: The Revised Oxford Translation.* Vol. 2, edited by J. Barnes, 1986–2129. Princeton, NJ: Princeton University Press.

_____. 2009. *Nicomachean Ethics.* Translated by Roger Crisp. Oxford: Oxford University Press.

Ashby, William Ross. 2012. *An Introduction to Cybernetics.* London: FIliquarian Legacy Publishing.

Augustine. 1960. *Confessions.* Translated by John K. Ryan. New York: Doubleday.

Aydin, Ciano. 2015. "The Artifactual Mind: Overcoming the 'Inside–Outside' Dualism in the Extended Mind Thesis and Recognizing the Technological Dimension of Cognition." *Phenomenology and the Cognitive Sciences* 14(1): 73–94.

Baier, Annette. 1986. "Trust and Antitrust." *Ethics* 96(2): 231–60.

_____. 1991. "Trust and Its Vulnerabilities" and "Sustaining Trust." *Tanner Lectures on Human Values.* Vol. 13. Salt Lake City: University of Utah Press.

Baker, Peter, and Julie Hirschfeld Davis. "Obama Chooses Nike Headquarters to Make His Pitch on Trade." *The New York Times*, May 7, 2015: http://www.nytimes.com/2015/05/08/business/obama-chooses-nike-headquarters-to-make-his-pitch-on-trade.html.

Baltimore, David, Paul Berg, Michael Botchan, Dana Carroll, R. Alta Charo, George Church, Jacob E. Corn, George Q. Daley, Jennifer A. Doudna, Marsha Fenner, Henry T. Greely, Martin Jinek, G. Steven Martin, Deward Penhoet, Samuel H. Sternberg, Jonathan S. Weissman, and Keith R. Yamamoto. 2015. "A Prudent Path Forward for Genomic Engineering and Germline Modification." *Science* 348(6230): 34–36.

Balzac, Honoré de. 1977. *The Wild Ass's Skin.* London: Penguin.

Barthes, Roland. 1978. "Death of the Author." Translated by S. Heath. In *Image, Music, Text*, 142–49. New York: Hill and Wang.

Benjamin, Walter. 2006. "The Work of Art in the Age of Its Technological Reproducibility." Translated by Harry Zohn. In *Walter Benjamin: Selected Writings, Volume 4, 1938–1940,* edited by Howard Eiland and Michael W. Jennings, 251–83. Cambridge, MA: Belknap Press of Harvard University Press.

Bernard, Cohen I. 1985. *The Birth of a New Physics.* New York: W. W. Norton.

Bijker, Wiebe E., Thomas P. Hughes, and Trevor Pinch, eds. 1993. *The Social Construction of Technological Systems: New Directions in the Sociology and History of Technology.* Cambridge, MA: MIT Press.

Black Mirror. 2011. "The Entire History of You." Directed by Brian Wilson. Written by Jesse Armstrong. BBC.

Bohman, James. 2008. "The Transformative of the Public Sphere: Political Authority, Communicative Freedom, and Internet Publics." In *Information Technology and Moral Philosophy*, edited by M. J. van den Joven and J. Weckert, 66–92. Cambridge: Cambridge University Press.

Bolzano, Bernard. 1837/1972. *Theory of science, attempt at a detailed and in the main novel exposition of logic with constant attention to earlier authors.* Edited and translated by Rolf George. Berkeley: University of California Press.

Borgmann, Albert. 1984. *Technology and the Character of Contemporary Life: A Philosophical Inquiry.* Chicago: University of Chicago Press.

Box, Richard C. 2007. "Redescribing the Public Interest." *The Social Science Journal* 44: 585–98.

Brabham, Daren C. 2013. *Crowdsourcing*. Cambridge, MA: MIT Press.

Bray, Abigail. 2009. "Chemical-Control: From the Cane to the Pill." In *Deleuze and New Technology*, edited by D. Savat and M. Poster, 82–103. Edinburgh, Scotland: Edinburgh University Press.

Brey, Philip. 2008. "Technological Construction of Power." *Social Epistemology: A Journal of Knowledge, Culture and Policy* 22(1): 71–95.

_____. 2010. "Philosophy of Technology after the Empirical Turn." *Techné: Research in Philosophy and Technology* 14(1): 36–48.

_____. 2012. "Anticipatory Ethics for Emerging Technologies." *Nanoethics* 6: 1–13.

Brown, Mark B. 2007. "Can Technologies Represent their Publics?" *Technology in Society* 29: 327–38.

Browning, Douglas. 1990. "Some Meanings of Automobiles." In *Technology as a Human Affair*, edited by L. A. Hickman, 172–77. New York: McGraw Hill.

Bryan, Nick Robert, ed. 2010. *Introduction to the Science of Medical Imaging*. New York: Cambridge University Press.

Busch, Lawrence. 2011. *Standards: Recipes for Reality*. Cambridge, MA: MIT Press.

Callon, Michel. 1993. "Society in the Making: The Study of Technology as a Tool for Social Analysis." In *The Social Construction of Technological Systems: New Directions in the Sociology and History of Technology*, edited by E. W. Bijker, T. P. Hughes, and T. Pinch, 83–103. Cambridge, MA: MIT Press.

Campbell, John. 1997. "The Structure of Time in Autobiographical Memory." *European Journal of Philosophy* 5: 105–18.

Carr, Nicholas. 2010. *The Shallows*. New York: W.W. Norton & Company.

Carson, Rachel. 1962. *Silent Spring*. New York: Houghton Mifflin.

Carter, Rita. 2010. *Mapping the Mind*. New York: Phoenix Paperback.

Chalmers, David, and Andy Clark. 1998. "The Extended Mind." *Analysis* 58(1): 7–19.

Chartier, Anne-Marie. 2008. "Cultural Perspective on Literacy Teaching and Methods for Young Readers." *Paedagogica Historica* 44(1/2): 7–29.

Chaulk, K., and T. Jones. 2011. "Online Obsessive Relational Intrusion: Further Concerns About Facebook." *Journal of Family Violence* 26: 245–54.

Chun, Wendy Hui Kyong. 2006. *Control and Freedom: Power and Paranoia in the Age of Fiber Optics*. Cambridge, MA: MIT Press.

Clark, Andy. 2010a. "Material Surrogacy and the Supernatural: Reflections on the Role of Artefacts in 'Off-Line' Cognition." In *The Cognitive Life of Things: Recasting the Boundaries of the Mind*, edited by Lambros Malafouris and Colin Renfrew, 23–27. Cambridge: McDonald Institute for Archaeological Research.

_____. 2010b. *Supersizing the Mind: Embodiment, Action, and Cognitive Extension*. Oxford: Oxford University Press.

Clayton, Russell. B., Randall E. Osborne, and Brian K. Miller. 2013. "Loneliness, Anxiousness, and Substance Use as Predictors of Facebook Use." *Computers in Human Behavior* 29: 687–93.

Coady, Cecil Anthony John. 1992. *Testimony: A Philosophical Study*. Oxford: Oxford University Press.

Coady, David. 2012. *What to Believe Now: Applying Epistemology to Contemporary Issues*. Malden, MA: Wiley-Blackwell.

Coeckelberg, Mark. 2013. "Human Being @ Risk." *Enhancement, Technology, and the Evaluation of Vulnerability Transformations*. Netherlands: University of Delft.

Collingridge, David. 1980. *The Social Control of Technology*. London: Frances Pinter.

Cook, Philip, and Conrad Heilman. 2010. "Censorship and Two Types of Self-Censorship." *LSE Choice Group*. The Centre for Philosophy of Natural and Social Science 6(2): 1–5.

_____. 2013. "Two Types of Self-Censorship: Public and Private." *Political Studies* 61: 178–96.

Cowen, Ruth S. 1983. *More Work for Mother: The Ironies of Household Technology from the Open Hearth to the Microwave*. New York: Basic Books.

Cowen, T., and A. Tabarrok. 2015. "The End of Asymmetric Information." http://www.cato-unbound.org/2015/04/06/alex-tabarrok-tyler-cowen/end-asymmetric-information. Accessed 04/10/2015.

Cravens, J. D., K. R. Leckie, and J. B.Whiting. 2013. "Facebook Infidelity: When Poking Becomes Problematic." *Contemporary Family Therapy* 35: 74–90.

Crowell, Steven. 2006/2009. "Husserlian Phenomenology." In *A Companion to Phenomenology and Existentialism*, edited by H. L. Dreyfus and M. A. Wrathall, 9–30. Blackwell Publishing Ltd.

Danaher, J. 2014. "Rule by Algorithm? Big Data and the Threat of Algocracy." http://ieet.org/index.php/IEET/more/danaher20140107. Accessed 02/23/2015.

Dawkins, Richard. 1989. *The Selfish Gene*. Oxford: Oxford University Press.

Dean, Mitchell. 2010. *Governmentality: Power and Rule in Modern Society*. London: Sage Publications Ltd.

Dechert, Charles R., eds. 1966. *The Social Impact of Cybernetics*. South Bend, IN: University of Notre Dame Press.

Deleuze, Gilles. 1986. *Cinema 1: Movement-Image*. Translated by H. Tomlinson and B. Habberjam. Minneapolis: University of Minnesota Press.

_____. 1992. "Postscript on the Societies of Control." *October* 59: 3–7.

_____. 2002. *Francis Bacon: The Logic of Sensation*. Translated by Daniel W. Smith. Minneapolis: University of Minnesota Press.

Deleuze, Gilles, and Felix Guattari. 1987. *A Thousand Plateaus*. Translated by B. Massumi. *Capitalism & Schizophrenia. Vol. II*. Minneapolis: University of Minnesota Press.

Denker, Alfred, ed. 2015. *Heidedgger und die technische Welt: Heidegegr Jahrbuch 9*. Freiburg: Alber.

Derrida, Jacques. 1978. "Structure, Sign, and Play." Translated by A. Bass. In *Writing and Difference*. Chicago: University of Chicago.

Dewey, John. 1997. *Experience and Education*. New York: Touchstone.

Donald, Merlin. 1991. *Origins of the Modern Mind*. Cambridge, MA: Harvard University Press.

Donath, J., and D. Boyd. 2004. "Public Displays of Connection." *BT Technology Journal* 2(4): 71–82.

Dorpat, Theodore L. 1996. *Gaslighting, the Double Whammy, Interrogation, and Other Methods of Covert Control in Psychotherapy and Pscyhoanalysis*. New York: Jason Aronson.

Dorrestijn, Steven. 2012. "Theories and Figures of Technical Mediation." In *Design and Anthropology*, edited by J. Donovan and W. Gunn, 219–30. Burlington, VT: Ashgate.

Drahos, P., and J. Braithwaite. 2002. *Information Feudalism.* London: Earthscan.

Dreyfus, Hubert L. 2001/2009. *On the Internet.* London: Routledge.

_____. 2013. "Anonymity versus Commitment: The Dangers of Education on the Internet." In *Philosophy of Technology: The Technological Condition*, edited by R. Scharf, C. Robert, and V. Dusek, 641–47. Somerset, NJ: John Wiley & Sons, Inc.

Drouin, Michelle, Daren H. Kaiser, and Daniel A. Miller. 2012. "Phantom Vibrations Among Undergraduates: Prevalence and Associated Psychological Characteristics." *Computers in Human Behavior* 28(4): 1490–96.

Dunagan, J. 2014. "Who Owns the Extended Mind? The Neuropolitics of Intellectual Property Law." In *The Sage Handbook of Intellectual Property*, edited by M. David and D. Halbert, 689–707. London: Sage.

Dyer-Witheford, N. 1999. *Cyber-Marx.* Chicago: University of Chicago Press.

Ellis, Albert. 2002. "Is Psychoanalysis Harmful?" *Psychiatric Opinion* 5(1): 16–25.

Ellul, Jacques. 1964. *The Technological Society.* Translated by J. Wilkinson. New York: Vintage Books.

_____. 2000. "Ideas of Technology: The Technological Order." *New Perspectives Quarterly* 17(3): 21–30.

Emarketer. 2015. "2 Billion Consumers Worldwide to Get Smart(phones) by 2016." http://www.emarketer.com/Article/2-Billion-Consumers-Worldwide-Smartphones-by-2016/1011694. Accessed 04/15/2015.

Embree, Lester, ed. 2004. *Gurwitsch's Relevancy for Cognitive Science.* Dordrecht: Springer.

Encabo, Jesús Vega, and F. Javier Gil Martin. 2007. "Science as Public Sphere?" *Social Epistemology: A Journal of Knowledge, Culture and Policy* 21(1): 5–20.

Engelland, Chad. 2010. "Unmasking the Person." *International Philosophical Quarterly* 50: 447–60.

_____. 2014. *Ostension: Word Learning and the Embodied Mind.* Cambridge, MA: MIT Press.

Epictetus. 2005. *Epictetus' Handbook and the Tablet of Cebes.* Translated by Keith Seddon. New York: Routledge Press.

Ess, C., and H. Fossheim. 2013. "Personal Data: Changing Selves, Changing Privacies." In *Digital Enlightment Yearbook 2013*, edited by M. Hildebrandt et al. 40–55. Amsterdam: IOS Press.

Estlund, D. 2003. "Why Not Epistocracy?" In *Desire, Identity and Existence: Essays in honor of T.M. Penner*, 53–69. New York: Academic Printing and Publishing.

Fardouly, Jasmine, Phillippa C. Diedrichs, Lenny R. Vartanian, and Emma Halliwell. 2015. "Social Comparisons on Social Media: The Impact of Facebook on Young Women's Body Image Concerns and Mood." *Body Image* 13: 38–45.

Feenberg, Andrew. 1995. *Alternative Modernity: The Technical Turn in Philosophy and Social Theory.* Berkeley, CA: University of California Press.

_____. 1999. *Questioning Technology.* London: Routledge.

_____. 2002a. *Critical Theory of Technology.* Oxford: Oxford University Press.

_____. 2002b. *Transforming Technology: A Critical Theory Revisited.* New York: Oxford University Press.

Fellows, Jennifer Jill. 2014. "Downstream of the Experts: Trust-Building and the Case of MPAs." *Social Epistemology: A Journal of Knowledge, Culture and Policy* 28(2): 187–207.

Fichte, J. G. 1982. *The Science of Knowledge*. Translated by P. Heath and J. Lachs. Cambridge: Cambridge University Press.

Fish, Jefferson M. 2015. "A University Is Not Walmart." *Psychology Today* (February 3). https://www.psychologytoday.com/blog/looking-in-the-cultural-mirror/201502/university-is-not-walmart. Accessed 06/24/2015.

Fisher, Matthew, Mariel K. Goddu, and Frank C. Keil. 2015. "Searching for Explanations: How the Internet Inflates Estimates of Internal Knowledge." *Journal of Experimental Psychology: General*. doi:10.1037/xge0000070. Accessed 04/10/2015.

Floridi, L. 2014. *The Fourth Revolution: How the Infosphere is Reshaping Human Reality*. Oxford: Oxford University Press.

Forster, E. M. 1909. "The Machine Stops." *Oxford and Cambridge Review* (November 8): 83–122.

Foucault, Michel. 1970. "Theatrum Philosophicum." Translated by D. F. Brouchard and S. Simon. *Critique* 282: 885–908.

_____. 1978. *The History of Sexuality Volume 1: An Introduction*. Translated by R. Hurley. New York: Pantheon Books.

_____. 1994. *The Order of Things: An Archaeology of the Human Sciences*. Vintage Books. New York: Vintage Books.

_____. 1995. *Discipline and Punish: The Birth of the Prison*. New York: Vintage Books.

_____. 2001. *Madness and Civilization: A History of Insanity in the Age of Reason*. Translated by R. Howard. London: Routledge.

_____. 2002. *Archaeology of Knowledge*. Translated by A. M. Sheridan Smith. London: Routledge.

_____. 2003. *"Society Must Be Defended": Lectures at the Collège de France, 1975–76*. New York: Picador.

_____. 2004a. *Abnormal: Lectures at the Collège de France, 1974–1975*. New York: Picador.

_____. 2004b. *Security, Territory, Population*. Translated by G. Burchell. New York: Palgrave Macmillan.

_____. 2008. *The Birth of Biopolitics: Lectures at the Collège de France, 1978–1979*. Translated by G. Burchell. New York: Palgrave Macmillan.

_____. 2010. *The Government of Self and Others: Lectures at the Collège de France, 1982–1983*. Basingstoke: Palgrave Macmillan.

Fraser, N. 2014. "Transnationalizing the Public Sphere." In *Transnationalizing the Public Sphere*, edited by K. Nash, 8–42. Cambridge: Polity.

Fricker, M. 2009. *Epistemic Injustice: Power and the Ethics of Knowing*. Oxford: Oxford University Press.

Frissen, V. 2011. "De Transparante Samenleving." *Jaarboek ICT en Samenleving 2011*. Gorredijk: Media Update Vakpublicaties.

Fromm, Georg H., William Leiss, John David Ober, Arno Waserman, and Edward J. Wilkins. 1964. *Review of One-Dimensional Man*. http://www.nybooks.com/articles/archives/1964/mar/19/one-dimensional-man-2/.

Fuller, Matthew, and Andrew Goffey. 2012. *Evil Media*. Cambridge, MA: MIT Press.

Fuller, Steve. 1991. *Social Epistemology*. Bloomington, IN: Indiana University Press.

_____. 1993. *Philosophy of Science and Its Discontents* (Orig. 1989). New York: Guilford Press.

_____. 2000. *The Governance of Science*. Buckingham: Open University Press.

_____. 2002. *Knowledge Management Foundations*. Woburn, MA: Butterworth-Heinemann.

_____. 2006a. *The New Sociological Imagination*. London: Sage.

_____. 2006b. "The Market: Source or Target of Morality?" In *The Moralization of the Market*, edited by N. Stehr, C. Henning, and B. Weiler, 129–53. New Brunswick, NJ: Transaction Books.

_____. 2009. "Review of the *Handbook of Science and Technology Studies*." *Isis* 100(1): 207–9.

_____. 2010. "The New Behemoth: Review of *Acting in an Uncertain World*." *Contemporary Sociology* 39(5): 533–36.

_____. 2011. *Humanity 2.0: What it Means to be Human Past, Present and Future*. London: Palgrave Macmillan.

_____. 2014. "Social Epistemology: The Future of an Unfulfilled Promise." *Social Epistemology Review and Reply Collective* 3(7): 29–37. http://wp.me/p1Bfg0-1wG.

_____. 2015. *Knowledge: The Philosophical Quest in History*. London: Routledge.

Fuller, Steve, and James Collier. 2004. *Philosophy, Rhetoric, and the End of Knowledge: A New Beginning for Science and Technology Studies*. Mahwah, NJ: Lawrence Erlbaum Associates.

Fuller, Steve, and Veronika Lipinska. 2014. *The Proactionary Imperative: A Foundation for Transhumanism*. London: Palgrave Macmillan.

Gabriëls, René. 2001. *Intellectuelen in Nederland: publieke controversen over kernenergie, armoede en Rushdie*. Amsterdam: Boom.

Gadamer, Hans-Georg. 1996. *The Enigma of Health: The Art of Healing in a Scientific Age*. Translated by J. Gaiger and R. Walker. Oxford: Polity Press.

Gee, J. P. 2007. *What Video Games Have to Teach Us About Learning and Literacy*. New York: Palgrave/Macmillan.

Gehlen, A. 1940. *Der Mensch. Seine Natur und seine Stellung*. Berlin: Junker und Dünnhaupt.

Gershenfield, N. 2012. "How to Make Almost Anything: The Digital Fabrication Revolution." *Foreign Affairs*. http://www.foreignaffairs.com/articles/138154/neil-gershenfeld/how-to-make-almost-anything. Accessed 03/05/2015.

Gilbert, W. 1991. "Towards a Paradigm Shift in Biology." *Nature* (January 10) 349: 99.

Gilovich, Thomas. 1993. *How We Know What Isn't So*. New York: Free Press.

Gitelman, Lisa. 2004. *New Media, 1740–1915*. Cambridge, MA: MIT Press.

_____. 2006. *Always Already New: Media, History and the Data of Culture*. Cambridge, MA: MIT Press.

_____. 2014. *Paper Knowledge: Toward a Media History of Documents*. Durham: Duke University Press.

Gladwell, M. 2000. *The Tipping Point: How Little Things Can Make a Big Difference*. New York: Little, Brown and Company.

Gleick, James. 2011. *The Information: A History, A Theory, A Flood.* New York: Random House.

Goffman, Erving. 1959. *The Presentation of Self in Everyday Life*. New York, NY: Anchor Books.

Goldberg, Sanford C. 2013. "'Analytic Social Epistemology' and the Epistemic Significance of Other Minds." *Social Epistemology Review and Reply Collective* 2(8): 26–48. http://social-epistemology.com/2013/07/10/analytic-social-epistemology-and-the-epistemic-significance-of-other-minds-sanford-c-goldberg/.

Goldman, Alvin. 2008. "The Social Epistemology of Blogging." In *Information Technology and Moral Philosophy*, edited by Jeroen van den Hoven and John Weckert. Cambridge, UK: Cambridge University Press.

Goody, J., and I. Watt. 1968. "The Consequences of Literacy." In *Literacy in Traditional Societies*, edited by J. Goody, 27–68. Cambridge: Cambridge University Press.

Gordon, Jon W. 1999. "Genetic Enhancements in Humans." *Science* 283: 2023–24.

Gould, Jeff. 2014. "The Natural History of Gmail Data Mining." *Medium.* https://medium.com/@jeffgould/the-natural-history-of-gmail-data-mining-be115d196b10. Accessed 04/21/2015.

Govier, T. 1992. "Trust, Distrust, and Feminist Theory." *Hypatia: A Journal of Feminist Philosophy* 7(1): 16–33.

_____. 1993. "Self-Trust, Autonomy, and Self-Esteem." *Hypatia: A Journal of Feminist Philosophy* 8(1): 99–120.

Gramsci, Antonio. 1971. *Selections From the Prison Notebooks*. Translated and edited by Q. Horare and G. N. Smith. London: Lawrence and Wishart.

Green, Ronald M. 2007. *Babies by Design: The Ethics of Genetic Choice*. New Haven: Yale University Press.

Groves, C., R. Lee, L. Frater, and G. Harper. 2010. *CSR in the UK Nanotechnology Industry: Attitudes and Prospects*. Working Paper Series No. 57, July 1. Cardiff: Centre for Business Relationships, Accountability, Sustainability and Society.

Guillebaud, Jean-Claude. 1999. *The Tyranny of Pleasure*. New York, NY: Algora Publishing.

Gurwitsch, A. 1964. *Field of Consciousness*. Pittsburgh, PA: Dusquesne University Press.

Guston, David H. 1993. "The Essential Tension in Science and Democracy." *Social Epistemology: A Journal of Knowledge, Culture and Policy* 7(1): 3–23.

_____. 2008. "Preface." In *The Yearbook of Nanotechnology in Society: Presenting Futures*. Vol. 1, edited by E. Fisher et al. v–viii. New York: Springer.

_____. 2014. "Understanding 'Anticipatory Governance.'" *Social Studies of Science* 44(2): 218–42.

Gutiérrez-Maldonado, José, Marta Ferrer-Garcia, Alejandra Caqueo-Urizar, and Elena Moreno. 2010. "Body Image in Eating Disorders: The Influence of Exposure to Virtual-Reality Environments." *Cyberpsychology, Behavior, and Social Networking* 13: 521–31.

Habermas, Jürgen. 1998. "Three Normative Models of Democracy." In *The Inclusion of the Other: Studies in Political Theory*, edited by J. Habermas, 239–52. Cambridge, MA: MIT Press.

Hanson, F. Allan. 2004. "From Classification to Indexing: How Automation Transforms the Way We Think." *Social Epistemology: A Journal of Knowledge, Culture and Policy* 18(4): 334–56.

Hardin, G. 1968. "The Tragedy of the Commons." *Science* 162: 1243–48.

Hardin, R. 2002. *Trust and Trustworthiness*. New York: Russell Sage Foundation.

Hardt, Michael, and Antonio Negri. 2001. *Empire*. Cambridge, MA: Harvard University Press.

Hayek, F. A. 2006. *The Constitution of Liberty*. London: Routledge.

Hayles, Katherine N. 2007. "Hyper and Deep Attention: The Generational Divide in Cognitive Modes." *Profession 2007*, 187–99. New York: Modern Language Association.

Heffernan, Virginia. 2016. *Magic and Loss: The Pleasures of the Internet*. New York: Simon & Schuster.

Hegel, G. W. F. 2008. *Outlines of the Philosophy of Right*. Translated by T. M. Knox. Oxford: Oxford University Press.

Heidegger, Martin. 1962. *Being and Time*. Translated by J. Macquarrie and E. Robinson. New York: Harper & Row.

_____. 1966. *Discourse on Thinking*. Translated by J. M. Anderson and E. H. Freund. New York: Harper & Row (DT).

_____. 1973. *The End of Philosophy*. Translated by J. Stambaugh. New York: Harper & Row (EP).

_____. 1976. "Nur noch ein Gott kann uns retten." *Der Spiegel* 31(May): 193–219. Translated by W. Richardson as "Only a God Can Save Us," in 1981. *Heidegger: The Man and the Thinker*, edited by T. Sheehan, 45–67, New York: Transaction Publishers.

_____. 1983. *Einführung in die Metaphysik (Summer semester 1935)*, edited by P. Jaeger. Frankfurt am Main: Vittorio Klostermann (GA 40).

_____. 1994. *Beiträge zur Philosophie (Vom Ereignis) (1936–1938)*, edited by F.-W. von Herrmann. Frankfurt am Main: Vittorio Klostermann (GA 65).

_____. 1996a. *Being and Time*. Translated by J. Stambaugh. New York: SUNY Press.

_____. 1996b. *Erläuterungen zu Hölderlins Dichtung (1936–1968)*, edited by F. W. von Herrmann, Frankfurt am Main: Vittorio Klostermann (GA 4).

_____. 1997. *Besinnung (1938/39)*, edited by F.-W. von Herrmann. Frankfurt am Main: Vittorio Klostermann (GA 66).

_____. 1999a. *Hölderlins Hymnen "Germanien" und "Der Rhein" (Winter semester 1934/35)*, edited by S. Ziegler. Frankfurt am Main: Vittorio Klostermann (GA 39/ HHGR).

_____. 1999b. *Metaphysik und Nihilismus*, edited by H.-J. Friedrich. Frankfurt am Main: Vittorio Klostermann (GA 67).

_____. 2000a. *An Introduction to Metaphysics*. Translated by G. Fried and R. Polt. New Haven and London: Yale University Press (IM).

_____. 2000b. *Reden und andere Zeugnisse eines Lebensweges (1910–1976)*, edited by H. Heidegger. Frankfurt am Main: Vittorio Klostermann (GA 16).

_____. 2000c. *Vorträge und Aufsätze (1936–1953)*, edited by F.-W. von Herrmann. Frankfurt am Main: Vittorio Klostermann (GA 7).

_____. 2001. *Poetry, Language, Thought*. Translated by A. Hofstadter. New York: Harper Collins (PLT).

_____. 2002. *Off the Beaten Track*. Translated by J. Young and K. Haynes. Cambridge: Cambridge University Press (OBT).

_____. 2003. *Holzwege (1935–1946)*, edited by F.-W. von Herrmann. Frankfurt am Main: Vittorio Klostermann (GA 5).

———. 2004. *Nature from Within*. Pittsburgh: University of Pittsburgh Press.

———. 2006. *Identität und Differenz (1955–1957)*, edited by F.-W. von Herrmann. Frankfurt am Main: Vittorio Klostermann (GA 11).

———. 2008. "The Question Concerning Technology." In *Basic Writings*, edited by D. F. Krell, 307–43. London: Harper & Row Perennials (QCT).

Heidelberger, Michael. 2004. *Nature from Within: Gustav Theodor Fechner and His Psychophysical Worldview*. Pittsburgh, PA: University of Pittsburgh Press.

Henty, G. A. 1886. *Through the Fray: A Tale of the Luddite Riots*. London: Blackie.

Heraclitus. 1979. *The Art and Thought of Heraclitus*. Translated by Charles Kahn. Cambridge: Cambridge University Press.

Hermberg, K. 2006. *Husserl's Phenomenology: Knowledge, Objectivity and Others*. London: Continuum International.

Hertlein, K. M., and F. P. Piercy. 2006. "Internet Infidelity: A Critical Review of the Literature." *The Family Journal of Counseling and Therapy for Couples and Families* 14: 366–71.

Hesper, E. J., and M. T. Whitty. 2010. "Netiquette Within Married Couples: Agreement About Acceptable Online Behavior and Surveillance Between Partners." *Computers in Human Behavior* 26: 919–26.

Hilkins, James, and Ricard Cherwitz. 2011. "On the Ontological and Epistemological Dimensions of Expertise: Why 'Reality' and 'Truth' Matter and How We Might Find Them." *Social Epistemology: A Journal of Knowledge, Culture and Policy* 25(3): 291–308.

Hill, B. M. 2014. "Google Has Most of My Email Because It Has All of Yours." http://mako.cc/copyrighteous/google-has-most-of-my-email-because-it-has-all-of-yours. Accessed 03/02/2015.

Hinchman, E. 2003. "Trust and Diachronic Agency." *Nous* 37(1): 23–51.

Hirst, William, and Gerald Echterhoff. 2012. "Remembering in Conversations: The Social Sharing and Reshaping of Memories." *Annual Review of Psychology* 63: 55–69.

Ho, A. 2011. "Trusting Experts and Epistemic Humility in Disability." *International Journal of Feminist Approaches to Bioethics* 4(2): 102–23.

Hollis, M., and S. Lukes, eds. 1982. *Rationality and Relativism*. Cambridge, MA: MIT Press.

Hopkins, Patrick D. 2009. "The Intersection of Culture, Gender, and Technology." In *Technology and Society: Building our Sociotechnical Future*, edited by D. G. Johnson and J. M. Wetmore, 195–203. Cambridge, MA: MIT Press.

Hughes, Thomas P. 1983. *Networks of Power: Electrification in Western Society, 1880–1930*. Baltimore: Johns Hopkins University Press.

———. 1987. "The Evolution of Large Technical Systems." In *The Social Construction of Technological Systems*, edited by W. E. Bijker, T. P. Hughes, and T. Pinch, 51–82. Cambridge, MA: MIT Press.

Huhtamo, Erkki. 2011. *Media Archaeology: Approaches, Applications, and Implications*. Berkeley, CA: University of California Press.

Husserl, Edmund. 1901/2001. *Logical Investigations*, Vol. I. Translated by D. Moran. New York: Routledge.

———. 1931. *Ideas: A General Introduction to Pure Phenomenology*. Translated by W. R. B. Gibson. New York, NY: HarperCollins Publishers Ltd.

_____. 1960. *Cartesian Meditations. An Introduction to Phenomenology*. Translated by D. Cairns. Boston, MA: Martinus Nijhoff Publishers.

_____. 2005. *Phantasy, Image Consciousness, and Memory (1898–1925)*. Translated by J. B. Brough. New York, NY: Springer.

_____. 2006. *The Basic Problems of Phenomenology (from the Lectures, Winter Semester 1910–1911)*. Translated by I. Farin and J. G. Hart. New York: Springer.

Hutchins, Edwin. 1995. *Cognition in the Wild*. Cambridge, MA: MIT Press.

Ihde, Don. 1993. *Postphenomenology*. Evanston, IL: Northwestern University Press.

Jaggar, Alison. 1991. "Feminist Ethics: An Introduction." *Forum: A Women's Studies Periodical* 16: 1.

Jefferson, Thomas. 1779. "Bill Establishing Religious Freedom." *Founders Online*. http://founders.archives.gov/documents/Jefferson/01-02-02-0132-0004-0082.

Jeffreys, Sheila. 2008. *The Industrial Vagina: The Political Economy of the Global Sex Trade*. London: Routledge.

Jelenchick, Lauren A., Jens C. Eickhoff, and Megan A. Moreno. 2013. "'Facebook Depression?' Social Networking Site Use and Depression in Older Adolescents." *Journal of Adolescent Health* 52(1): 128–30.

Johnson, Dave. 2015. "Obama to Visit Nike to Promote TPP. Wait, *Nike*? Really?" *Huffington Post*, May 7, 2015: http://www.huffingtonpost.com/dave-johnson/obama-to-visit-nike-to-pr_b_7233118.html.

Johnson, Deborah G. 2011. "Software Agents, Anticipatory Ethics and Accountability." In *The Growing Gap Between Emerging Technologies and Legal and Ethical Oversight: The Pacing Problem*, edited by G. E. Marchant et al., 61–76. New York: Springer Science+Business B.V.

Johnston, John. 2008. *The Allure of Machinic Life: Cybernetics, Artificial Life, and the New AI*. London: MIT Press.

Johnston, Ron. 1984. "Controlling Technology: An Issue for the Social Studies of Science." *Social Studies of Science* 14: 97–133.

Jonas, Hans. 1953. A Critique of Cybernetics. *Social Research* 20(1): 172–192.

Jones, Karen. 1996. "Trust as an Affective Attitude." *Ethics* 107: 4–25.

Jung, C. G. 1910. "The Association Method." *American Journal of Psychology* 31: 219–69.

Jünger, Ernst. 1963/1983. "Technology as the Mobilization of the World through the *Gestalt* of the Worker." Translated by J. M. Vincent, revised by R. J. Kundell. In *Philosophy and Technology: Readings in the Philosophical Problems of Technology*, edited by C. Mitchum and R. Mackey, 269–89. New York: The Free Press.

Kahneman, Daniel. 2013. *Thinking, Fast and Slow*. New York: Farrar, Straus and Giroux.

Kaijser, Arne. 2003. "Redirecting Infrasystems Towards Sustainability. What can We Learn From History?" In *Individual and Structural Determinants of Environmental Practice*, edited by A. Biel, B. Hansson, and M. Mårtensson, 152–79. Aldershot: Ashgate.

Kanjo, Eiman. 2010. "NoiseSPY: A Real-Time Mobile Phone Platform for Urban Noise Monitoring and Mapping." *Mobile Networks and Applications* 15(4): 562–74.

Kant, Immanuel. 1784/1991. "An Answer to the Question: 'What is Enlightenment?'" In *Kant: Political Writings*. Translated by H. B. Nisbet. 54–60. Cambridge: Cambridge University Press.

_____. 1960. *Religion Within the Limits of Reason Alone*. Translated by T. M. Greene and H. H. Hudson. New York: Harper & Row.

_____. 1982. *Ausgewählte Schriften zur Pädagogik und ihrer Begründung*, edited by Hans-Hermann Groothoff. Paderborn: Ferdinand Schöningh Verlag.

_____. 1996. *Practical Philosophy*. Translated by Mary J. Gregor. Cambridge: Cambridge University Press.

_____. 1998. *Critique of Pure Reason*. Translated by P. Guyer and A. Wood. Cambridge: Cambridge University Press.

_____. 2007. *Anthropology, History, and Education*. Translated by R. B. Louden and G. Zöller. Cambridge: Cambridge University Press.

Karsten, Harries, ed. 1994. *Martin Heidegger: Politics, Art, and Technology*. New York: Holmes & Meier.

Kass, Leon. 1999. *The Hungry Soul: Eating and the Perfecting of Our Nature*. Chicago: University of Chicago Press.

_____. 2002. "The Problem of Technology and Liberal Technology." In *Life Liberty and the Defense of Dignity: The Challenge for Bioethics*. San Francisco: Encounter Books.

Kaufman, Leslie. 2013. "*Bombings Trip Up Reddit in Its Turn in Spotlight*." *New York Times*. April 28. http://www.nytimes.com/2013/04/29/business/media/bombings-trip-up-reddit-in-its-turn-in-spotlight.html. Accessed 03/15/2015.

Keiper, Adam. 2007. "Nanoethics as a Discipline?" *New Atlantis* (Spring): 55–67.

Keller, Evelyn F. 1995a. *Reflections on Gender and Science*. New Haven and London: Yale University Press.

_____. 1995b. "The Origin, History and Politics of the Subject Called 'Gender and Science': A First Personal Account." In *Handbook of Science and Technology Studies*, edited by S. Jasanoff et al., 80–94. Thousand Oaks: Sage Publications.

Kelly, Kevin. 2010. *What Technology Wants*. Viking: Penguin Group.

_____. 2014. "Why You Should Embrace Surveillance, Not Fight It." *Wired* (March 10).

Kierkegaard, Søren. 1962. *The Present Age: And of the Difference Between a Genius and an Apostle*. New York: Harper & Row.

_____. 1970. *Kierkegaard's Journals and Papers*. Translated by H. V. Hong and E. H. Hong. Bloomington: Indiana University Press.

_____. 1996. *Papers and Journals: A Selection*, edited by Alastair Hannay. New York: Penguin Books.

Kisiel, Theodore. 2014. "Heidegger and Our Twenty-First Century Experience of Ge-Stell." In *The Multidimensionality of Hermeneutic Phenomenology*, edited by B. Babich and D. Ginev, 137–52. New York: Springer.

Knöpfel, Thomas, and Edward Boyden, eds. 2012. *Optogenetics: Tools for Controlling and Monitoring Neuronal Activity*. New York: Elsevier.

Kockelkoren, P. 2014. "The Quest for the Sources of Self, Seen from the Vantage Point of Plessner's Material a Priori." In *Plessner's Philosophical Anthropology: Perspectives and Prospects*, edited by J. de Mul, 179–86. Amsterdam/Chicago: Amsterdam University Press/Chicago University Press.

Krimsky, Sheldon. 2003. *Science in the Private Interest*. Lanham, MD: Rowman and Littlefield.

Kroon Van Diest, A. and M. Perez. 2013. "Exploring the Integration of Thin-ideal Internalization and Self-objectification in the Prevention of Eating Disorders." *Body Image* 10: 16–25.

Kurzweil, Ray. 2005. *The Singularity is Near: When Humans Transcend Biology*. New York: Penguin Books.

Lanier, Jaron. 2011. *You are Not a Gadget: A Manifesto*. New York: Vintage Books.

Latour, Bruno. 1994. "On Technical Mediation—Philosophy, Sociology, Genealogy." *Common Knowledge* 3(2): 29–64.

_____. 1999. "A Collective of Humans and Nonhumans." In *Pandora's Hope: Essays on the Reality of Science Studies*, 198–215. Cambridge, MA: Harvard University Press.

_____. 2007. *Reassembling the Social: An Introduction to Actor-Network-Theory*. Oxford: Oxford University Press.

Latour, Bruno, and Steve Woolgar. 1986. *Laboratory Life: The Construction of Scientific Facts*. Princeton, NJ: Princeton University Press.

Leder, Drew. 1990. *The Absent Body*. Chicago: University of Chicago Press.

Lemke, Thomas. 2011. *Biopolitics: An Advanced Introduction*. New York: New York University Press.

Lessig, Lawrence. 2005. *Free Culture: The Nature and Future of Creativity*. New York: Penguin.

Levinas, Emmanuel. 1998a. "Dialogue on Thinking-of-the-Other." Translated by M. B. Smith and B. Harshav. In *Entre Nous*, edited by M. B. Smith and B. Harshav, 201–6. New York: Columbia University Press.

_____. 1998b. "Dying For . . ." Translated by M. B. Smith and B. Harshav. In *Entre Nous*, edited by M. B. Smith and B. Harshav, 207–19. New York: Columbia University Press.

_____. 1998c. "German Preface to *Totality and Infinity*." Translated by M. B. Smith and B. Harshav. In *Entre Nous*, edited by M. B. Smith and B. Harshav, 197–200. New York: Columbia University Press.

_____. 1998d. "Is Ontology Fundamental?" Translated by M. B. Smith and B. Harshav. In *Entre Nous*, edited by M. B. Smith and B. Harshav, 1–12. New York: Columbia University Press.

_____. 1998e. "The Other, Utopia, and Justice." Translated by M. B. Smith and B. Harshav. In *Entre Nous*, edited by M. B. Smith and B. Harshav, 223–33. New York: Columbia University Press.

Levy, Ariel. 2006. *Female Chauvinist Pigs: Women and the Rise of Raunch Culture*. New York: Free Press.

List, Christian, and Philip Pettit. 2010/2011. *Group Agency: The Possibility, Design, and Status of Corporate Agents*. Oxford: Oxford University Press.

Lohr, Steve. 2015. *Data-sim: The Revolution Transforming Decision Making, Consumer Behavior, and Almost Everything Else*. New York: Harper Collins.

Loscerbo, John. 1981. *Being and Technology:A Study in the Philosophy of Martin Heidegger*. The Hague: Martinus Nijhoff.

Losh, Elizabeth. 2014. "Hashtag Feminism and Twitter Activism in India." *Social Epistemology Review and Reply Collective* 3(3): 11–22. http://wp.me/p1Bfg0-1Kx.

Louvieris, P., D. Bell, N. Nomikos, and A. Guinchard. 2015. "Digital Prosumer: Establishing a Futures Market for Digital Personhood Data." (UK) Engineering and Physical Sciences Research Council Grant EP/L005859/1: http://www.digitalpersonhood.org/prosumer.php.

Lyotard, Jean-François. 1979/1984. *The Postmodern Condition: A Report on Knowledge*. Translated by G. Bennington and B. Massumi. Minneapolis: University of Minnesota Press.

Malafouris, Lambros. 2013. *How Things Shape the Mind.* Cambridge, MA: MIT Press.

Maguire, Eleanor A., Katherine Woollett, and Hugo J. Spiers. 2006. "London Taxi Drivers and Bus Drivers: A Structural MRI and Neuropsychological Analysis." *Hippocampus* 16(12): 1091–101.

Mangen, A., and J. Velay. 2010. "Digitizing Literacy: Reflections on the Haptics of Writing." In *Advances in Haptics*, edited by M. H. Zadeh. Intech, Open Science Publication: http://www.intechopen.com/books/advances-in-haptics/digitizing-literacy-reflections-on-the-haptics-of-writing. Accessed 03/05/2015.

Mann, Thomas. 1968/1949. *Doctor Faustus. The Life of the German Composer Adrian Leverkühn as told by a Friend.* London: Penguin Books.

Marcel, Gabriel. 1950. *The Mystery of Being, Volume I: Reflection and Mystery.* Translated by G. S. Fraser. South Bend: St. Augustine's Press.

_____. 1962. The Sacred in the Technological Age. *Theology Today* 19: 27–38.

Marcuse, H. 1991. *One-dimensional Man: Studies in the Ideology of Advanced Industrial Society.* Boston: Beacon Press.

Margulis, Lynn, and Dorion Sagan. 2007. *Dazzle Gradually: Reflections on the Nature of Nature.* White River Junction, VT: Chelsea Green Publishing.

Marshall, T. C., K. Bejanyan, G. Di Castro, and R. A.Lee. 2013. "Attachment Styles as Predictors of Facebook-related Jealousy and Surveillance in Romantic Relationships." *Personal Relationships* 20(1): 1–22.

Marx, Karl. 2005. *Das Kapital: Kritik der politischen Ökonomie*. Köln: Parkland.

Marx, Karl, and Michael Quante. 2009. *Ökonomisch-philosophische Manuskripte: Kommentar von Michael Quante*. Frankfurt a.M.: Suhrkamp.

Mayer-Schönberger, V., and K. Cukier. 2013. *Big Data: The Revolution that Will Transform How We Live, Work and Think.* London: John Murray.

McCloskey, D. 1975. "The Economics of Enclosure." In *European Peasants and their Markets*, edited by W. Parker and E. Jones, 123–60. Princeton, NJ: Princeton University Press.

McDonald, Aleecia M., and Lorrie F. Cranor. 2008. "The Cost of Reading Privacy Policies." *I/S: A Journal of Law and Policy for the Information Society* 4(3): 540–65.

McLeod, C. 2011. "Trust." *Stanford Encyclopedia of Philosophy*, edited by E. Zalta. http://plato.stanford.edu/archives/spr2011/entries/trust/.

McLuhan, Marshall. 1962. *The Gutenberg Galaxy.* Toronto: University of Toronto Press.

Medina, José. 2012. *The Epistemology of Resistance: Gender and Racial Oppression, Epistemic Injustice, and the Social Imagination.* Oxford: Oxford University Press.

Merleau-Ponty, M. 2012. *Phenomenology of Perception.* Translated by Donald A. Landes. New York: Routledge.

Merriam-Webster.com. 2015. http://www.merriam-webster.com. Accessed 06/13/2015.

Michelfelder, Diane. 2001. "The Moral Value of Informational Privacy in Cyberspace." *Ethics and Information Technology* 3(2): 129–35.

_____. 2011. "Dirty Hands, Speculative Minds, and Smart Machines." *Philosophy and Technology* 24(1): 55–68.

Miller, G. 2013. "What Should We be Worried About? Chinese Eugenics." *Edge* (Question of the Year). https://edge.org/response-detail/23838.

Miller, Peter, and Nikolas Rose. 2008. *Governing the Present: Administering Economic, Social and Personal Life*. Cambridge: Polity Press.

Mitcham, Carl. 1994. *Thinking through Technology: The Path between Engineering and Philosophy*. Chicago, IL: University of Chicago Press.

_____. 1995. Notes toward a philosophy of meta-technology. *Techné* 1(2&3): 3–5.

Mitchell, Tom. 1997. *Machine Learning*. New York: McGraw Hill.

Mollenhauer, K. 2013. *Forgotten Connections. On Culture and Upbringing*. Translated by N. Friesen. London: Routledge.

Moriarty, Philipp. 2011. "Science as a Public Good." In *A Manifesto for the Public University*, edited by J. Holmwood, 56–73. London: Bloomsbury.

Morozov, Evgeny. 2012. *The Net Delusion: How Not to Liberate the World*. New York: Penguin Books.

_____. 2013a. "The Internet Ideology: Why We Are Allowed to Hate Silicon Valley." *Frankfurter Allgemeine Zeitung*. http://www.faz.net/aktuell/feuilleton/debatten/the-internet-ideology-why-we-are-allowed-to-hate-silicon-valley-12658406.html.

_____. 2013b. "The Real Privacy Problem." http://www.technologyreview.com/featuredstory/520426/the-real-privacy-problem/.

_____. 2015a. "Facebook isn't a Charity: The Poor Will Pay by Surrendering their Data." *Guardian* (London). April 26.

_____. 2015b. "Where Uber and Amazon Rule: Welcome to the World of the Platform." *Guardian* (London). June 7.

Mullins, Eustace. 2009. *The Secrets of the Federal Reserve*. New York: Bridger House Publishers.

Munn, Nicholas John. 2012. "The New Political Blogosphere." *Social Epistemology: A Journal of Knowledge, Culture and Policy* 26(1): 55–70.

NameTag. 2015. "NameTag App." http://www.nametag.ws. Accessed 04/20/2015.

Nietzsche, Friedrich. 1980. *On the Advantage and Disadvantage of History for Life*. Translated by Peter Preuss. Indianapolis: Hackett Publishing.

Nissenbaum, Helen. 2009. *Privacy in Context. Technology, Policy, and the Integrity of Social Life*. Stanford, CA: Stanford Law Books.

Norberg, P. A., and D. R. Horne. 2007. "The Privacy Paradox: Personal Information Disclosure Intentions versus Behaviors." *The Journal of Consumer Affairs* 41: 100–126.

Nordmann, Alfred. 2007. "If and Then: A Critique of Speculative Nanoethics." *Nanoethics* 1: 31–46.

_____. 2010. "A Forensics of Wishing: Technology Assessment in the Age of Technoscience." *Poiesis and Praxis* 7(1–2): 5–15.

Nordmann, Alfred, and Arie Rip. 2009. "Mind the Gap Revisited." *Nature Nanotechnology* 4(5): 273–74.

Norton, Bryan. 1999. "Ecology and Opportunity: Intergenerational Equity and Sustainable Options." In *Fairness and Futurity: Essays on Environmental Sustainability and Social Justice*, edited by A. Dobson, 118–50. Oxford: Oxford University Press.

Nozick, Robert. 1981. *Philosophical Explanations*. Cambridge MA: Harvard University Press.

Nussbaum, Martha. 1997. *Cultivating Humanity*. Cambridge, MA: Harvard University Press.

Offe, C. 2012. "Whose Good is the Common Good?" *Philosophy and Social Criticism* 38: 665–84.

Ohm, P. 2013. "Response: The Underwhelming Benefits of Big Data." *University of Pennsylvania Law Review Online* 339. http://www.pennlaw review.com/responses/8-2013/Ohm.pdf.

Olson, D. 1977. "From Utterance to Text: The Bias of Language in Speech and Writing." *Harvard Educational Review* 47: 257–81.

_____. 2006. "Response: Continuing the Discourse on Literacy." *Research in the Teaching of English* 41(2): 175–79.

Olson, D., and K. Oatley. 2014. "The Quotation Theory of Writing." *Written Communication* 31(1): 4–26.

Ong, W. 1982. *Orality and Literacy: The Technologizing of the Word*. London: Methuen.

Oudshoorn, Nelly. 1994. *Beyond The Natural Body: An Archeology of Sex Hormones*. New York: Routledge.

_____. 1999. "The Decline of the One-Size-Fits-All Paradigm, or, How Reproductive Scientists Try to Cope With Postmodernity." In *The Social Shaping of Technology*, edited by D. MacKenzie and J. Wajcman, 325–40. Buckingham, PA: Open University Press.

Owen, Richard, Phil Macnaghten, and Jack Stilgoe. 2012. "Responsible Research and Innovation: From Science in Society to Science for Society, with Society." *Science and Public Policy* 39(6): 751–60.

Ozdemir, V., D. Husereau, S. Hyland, S. Samper, and M. Salleh. 2009. "Personalized Medicine Beyond Genomics: New Technologies, Global Health Diplomacy and Anticipatory Governance." *Current Pharmacogenomics and Personalized Medicine* 7(4): 225–30.

Palen, Leysia, and Paul Dourish. 2003. "Unpacking 'Privacy' for a Networked World. CHI '03." *Proceedings of the SIGCHI Conference on Human Factors in Computing Systems*: 129–36.

Parikka, Jussi. 2010. "Cartographies of Media Archaeology." http://mediacartographies.blogspot.com/2010/10/what-is-media-archaeology-beta.html.

Pariser, Eli. 2011. *The Filter Bubble: What the Internet is Hiding From You*. New York: Penguin Press.

Perelman, Michael. 1991. *Information, Social Relations, and the Economics of High Technology*. New York: St. Martin's Press.

Pinch, Trevor J., and Wiebe Bijker. 1993. The Social Construction of Facts and Artifacts: Or How the Sociology of Science and the Sociology of Technology Might Benefit Each Other. In *The Social Construction of Technological Systems: New*

Directions in the Sociology and History of Technology, 11–44. Cambridge, MA: MIT Press.

Pinkard, Terry. 1986. "Freedom and Social Categories in Hegel's Ethics." *Philosophy and Phenomenological Research* 47(2): 209–32.

Plato. 1997a. "Apology." Translated by G. M. A. Grube. In *Plato: Complete Works*, edited by John M. Cooper, 17–36. Indianapolis: Hackett Publishing.

_____. 1997b. "Phaedrus." Translated by Alexander Nehamas and Paul Woodruff. *Plato Complete Works*, edited by John M. Cooper, 506–56. Cambridge: Hackett Publishing Company.

_____. 1997c. "Republic." Translated by G. M. A. Grube. Revised by C. D. C. Reeve. *Plato Complete Works*, edited by John M. Cooper, 971–1223. Cambridge: Hackett Publishing Company.

_____. 1997d. "Symposium." Translated by Alexander Nehamas and Paul Woodruff. In *Plato: Complete Works*, edited by John M. Cooper, 457–505. Indianapolis, IN: Hackett Publishing Company.

Plessner, Helmuth. 1975. *Die Stufen des Organischen und der Mensch*. Hawthorne, NY: Walter deGruyter.

_____. 1982. *Mit anderen Augen. Aspekte einer philosophischen Anthropologie*. Stuttgart: Reclam.

Plessner, H. Forthcoming. *The Levels of the Organic and Man*. Translated by Scott Davis.

Pogge, Thomas. 2001. "Priorities of Global Justice." *Metaphilosophy* 32: 6–24.

Porter, Theodore M. 1995. *Trust in Numbers*. Princeton, NJ: Princeton University Press.

Posner, Richard. 2005. "Bad News." *The New York Times*, July 31, 2005: 1–11.

Possidius. 1919. *Life of St. Augustine*. Translated by Herbert Theberath Weiskotten. Princeton, NJ: Princeton University Press.

Primack, Brian A., Brandi Swanier, Anna M. Georgiopoulos, Stephanie R. Land, and Michael J. Fine. 2009. "Association Between Media Use in Adolescence and Depression in Young Adulthood." *Archive of General Psychiatry* 66(2): 181–88.

Proctor, Robert. 1991. *Value-Free Science? Purity and Power in Modern Knowledge*. Cambridge, MA: Harvard University Press.

Pryzbylski, Andrew K., and Netta Winstein. 2012. "Can You Connect with Me Now? How the Presence of Mobile Communication Technology Influences Face-to-Face Conversation Quality." *Journal of Social and Personal Relationships* 30: 237–46.

Quay, Ray. 2010. "Anticipatory Governance: A Tool for Climate Change Adaptation." *Journal of the American Planning Association* 76(4): 496–511.

Radder, Hans. 1996. *In and About the World*. Albany: State University of New York Press.

_____. 2000. "Review Essay of 'The Governance of Science.'" *Science, Technology, and Human Values* 25: 520–26.

_____. 2008. "Critical Philosophy of Technology: The Basic Issues." *Social Epistemology: A Journal of Knowledge, Culture and Policy* 22(1): 51–70.

_____. 2010. "The Commodification of Academic Research." In *The Commodification of Academic Research: Science and the Modern University*, edited by H. Radder, 1–23. Pittsburgh: University of Pittsburgh Press.

Richardson, I. 2010. "Faces, Interfaces, Screens: Relational Ontologies of Framing, Attention and Distraction." *Transformations 18*. http://www.transformationsjournal.org/journal/issue_18/article_05.shtml. Accessed 03/05/2015.

Ricœur, P. 2007. *Husserl: An Analysis of His Phenomenology*. Evanston, IL: Northwestern University Press.

Roache, Rebecca. 2008. "Ethics, Speculation and Values." *Nanoethics* 2: 317–27.

Robinson, Bruce. 2015. "With a Different Marx: Value and the Contradictions of Web 2.0 Capitalism." *The Information Society* 31: 44–51.

Rosen, Jeffrey. 2012. "The Right to Be Forgotten." *Stanford Law Review Online* 64(88). http://www.stanfordlawreview.org/online/privacy-paradox/right-to-be-forgotten. Accessed 03/15/2015.

Rosen, Larry D., K. Whaling, S. Rab, L. Mark Carrier, and Nancy A. Cheever. 2013. "Is Facebook Creating 'iDisorders'? The Link Between Clinical Symptoms of Psychiatric Disorders and Technology Use, Attitudes and Anxiety." *Computers in Human Behavior* 29(3): 1243–54.

Rosenberg, Daniel, and Anthony Grafton. 2012. *Cartographies of Time*. New York: Princeton Architectural Press.

Rothschild, Emma. 2001. *Economic Sentiments: Adam Smith, Condorcet and the Enlightenment*. Cambridge, MA: Harvard University Press.

Roustang, François. 1982. *Psychoanalysis Never Lets Go*. Cleveland, OH: Johns Hopkins University Press.

_____. 1990. *The Lacanian Delusion*. Oxford: Oxford University Press.

Sakile, K. Camara, and Mark P. Orbe. 2011. "Understanding Interpersonal Manifestations of 'Reverse Discrimination' Through Phenomenological Inquiry." *Journal of Intercultural Communication Research* (June): 111–34.

Sappington, David E. M. 1991. "Incentives in Principal–Agent Relationships." *Journal of Economic Perspective* 5(2): 45–66.

Sartre, Jean-Paul. 1984. *Being and Nothingness*. Translated by Hazel E. Barnes. New York: Washington Square Press.

Sawyer, R. Keith. 2005. *Social Emergence: Societies As Complex Systems*. Cambridge: Cambridge University Press.

Sax, Joseph L. 1970. "The Public Trust Doctrine in Natural Resource Law: Effective Judicial Intervention." *Michigan Law Review* 68(3): 471–566.

Scalambrino, Francesco. 1998. *Un uomo sotto la Mole: Biografia di Antonio Gramsci*. Torino: Collana Bancarella.

Scalambrino, Frank. 2011. *Non-Being & Memory: A critique of pure difference in Derrida and Deleuze*. (Doctoral Dissertation). Retrieved from ProQuest. (UMI: 3466382).

_____. 2012. "Mnemo-psychography: The Origin of Mind and the Problem of Biological Memory Storage." In *Origins of Mind*, edited by L. Swan, series editors M. Barbieri and J. Hoffmeyer. *Biosemiotics (Vol. VIII)*, 327–39. New York: Springer.

_____. 2013a. "A Brief History of the Problem of Agent Causation in the Human and Behavioral Sciences with a Recommendation for Future Research." Paper presented at the 45th Annual Meeting of *Cheiron: The International Society for the History of Behavioral and Social Sciences*, Dallas, Texas, June 21–23.

_____. 2013b. "Implications of Gestalt Phenomenology's Kantian Roots for the Brain Localization Research Project." Paper presented at the 46th Annual Meeting of the *North Texas Philosophical Association*. Denton, Texas, April 4–6.

_____. 2014a. "From a Phenomenology of the Reciprocal Nature of Habits and Values to an Understanding of the Intersubjective Ground of Normative Social Reality." *Phenomenology and Mind* 6: 156–67.

_____. 2014b. "From a Statement of Its Vision Toward Thinking into the Desire of a Corporate Daimon." *Social Epistemology Review and Reply Collective* 3(10): 34–39.

_____. 2014c. "Philosophical Approaches to Technology." In *Philosophy of Technology: A Reader*, edited by F. Scalambrino, 1–20. San Diego, CA: Cognella.

_____. 2014d. "Social Media and the Cybernetic Mediation of Interpersonal Relations." In *Philosophy of Technology: A Reader*, edited by F. Scalambrino, 123–33. San Diego, CA: Cognella.

_____. 2015a. *Full Throttle Heart: Nietzsche, Beyond Either/Or*. New Philadelphia, OH: The Eleusinian Press.

_____. 2015b. Phenomenological Psychology. *Internet Encyclopedia of Philosophy*. http://www.iep.utm.edu/phen-psy/.

Schmidt, Eric. 2012. "Boston University 139th Commencement Address: Eric Schmidt." http://www.bu.edu/news/2012/05/20/boston-university-139th-commencement-address-eric-schmidt/. Accessed 03/05/2015.

Schmitz, Kenneth. 2007. "Geography of the Human Person." In *The Texture of Being: Essays in First Philosophy*, edited by Paul O'Herron. Washington, DC: The Catholic University of America Press.

Scoglio, S. 1998. *Transforming Privacy: A Transpersonal Philosophy of Rights*. Westport, CT: Greenwood Press.

Sconce, Jeffrey. 2000. *Haunted Media: Electronic Presence from Telegraphy to Television*. Durham, NC: Duke University Press.

Searle, John. 1995. *The Construction of Social Reality*. New York: Free Press.

Segelken, H. R., and S. Shackford. 2014. "Emotional contagion sweeps Facebook." *Cornell Chronicle* (Ithaca NY). June 10.

Selcer, Daniel. 2010. *Philosophy and the Book: Early Modern Figures of Material Inscription*. London: Bloomsbury Academic.

Serchuk, Adam. 1989. "What Can the Cognitive Psychology of Science Bring to Science and Technology Studies?" *Social Epistemology: A Journal of Knowledge, Culture and Policy* 3(2): 147–52.

Serrano, David Teira, and Jesús Zamora Bonilla. 2007. "Editors' Introduction: Science, Normativity and the Public." *Social Epistemology: A Journal of Knowledge, Culture and Policy* 21(1): 1–4.

Sismondo, Sergio. 2010. *An Introduction to Science and Technology Studies*. West Sussex: Blackwell Publishing.

Smith, Aaron. 2011. "Americans and Text Messaging." *Pew Research Center*. http://www.pewinternet.org/2011/09/19/americans-and-text-messaging/. Accessed 04/21/2015.

Smith, Adam. 2000. *The Wealth of Nations*. New York: Modern Library Random House.

Smith, H. J., T. Dinev and H. Xu. 2001. "Information Privacy Research: An Interdisciplinary Review." *MIS Quarterly* 35(4): 989–1015.

Sokolowski, Robert. 2000. *Introduction to Phenomenology*. Cambridge: Cambridge University Press.

Solove, J. Daniel. 2006. "A Taxonomy of Privacy." *University of Pennsylvania Law Review* 154(3): 477–560.

Spiegelberg, H. 1972. *Phenomenology in Psychology and Psychiatry: A Historical Introduction*. Evanston, IL: Northwestern University Press.

Spitzberg, B. H., and G. D. Hoobler. 2002. "Cyberstalking and the Technologies of Interpersonal Terrorism." *New Media and Society* 4: 71–92.

Stanley, J. 2005. *Knowledge and Practical Interests*. Oxford: Oxford University Press.

Steers, Mai-Ly N., Robert E. Wickham, and Linda K. Acitelli. 2014. "Seeing Everyone Else's Highlight Reels: How Facebook Usage is Linked to Depressive Symptoms." *Journal of Social and Clinical Psychology* 33(8): 701.

Steiner, C. 2012. *Automate This: How Algorithms Came to Rule the World*. London: Portfolio.

Sterckx, Sigrid, Julian Cockbain, Heidi C. Howard, and Pascal Borry. 2013. "'I Prefer a Child With . . .': Designer Babies, Another Controversial Patent in the Arena of Direct-to-Consumer Genomics." *Genetics in Medicine* 15: 923–24.

Sterne, Jonathan. 2003. *The Audible Past: Cultural Origins of Sound Reproduction*. Durham: Duke University Press.

———. 2012. *MP3: The Meaning of a Format*. Durham: Duke University Press.

Stiegler, B. 2008. *Prendre soin de la jeunesse et des générations*. Paris: Flammarion.

Stiegler, B., and I. Rogoff. 2010. "Transindividuation, E—flux, 14." http://www.e-flux.com/journal/transindividuation/. Accessed 03/05/2015.

Street, B. 2003. "What's 'New' in New Literacy Studies? Crtitical Approaches to Literacy in Theory and Practice." *Current Issues in Contemporary Education* 5(2): 77–91.

Strong, Richard Charles. 2014. "Habit, Distraction, Absorption: Reconsidering Walter Benjamin and the Relation of Architecture to Film." In *The Missed Encounter of Radical Philosophy with Architecture*, edited by Nadir Lahiji, 163–82. London: Bloomsbury.

Sunstein, C. 2014. *Why Nudge? The Politics of Libertarian Paternalism*. New Haven: Yale University Press.

Sutton, John. 2009. "Remembering." In *The Cambridge Handbook of Situated Cognition*, edited by P. Robbins and M. Aydede, 217–35. Cambridge: Cambridge University Press.

Sutton, John, Celia B. Harris, Paul G. Keil, and Amanda J. Barnier. 2010. "The Psychology of Memory, Extended Cognition, and Socially Distributed Remembering." *Phenomenology and the Cognitive Sciences* 9(4): 521–60.

Syn, Juliette. 2014. "The Social License: Empowering Communities and a Better Way Forward." *Social Epistemology: A Journal of Knowledge, Culture and Policy* 28(3–4): 318–39.

Szasz, Thomas. 2003. *Pharmacracy: Medicine and Politics in America*. New York: Syracuse University Press.

Thaler, Richard H., and Cass R. Sunstein. 2008/2009. *Nudge: Improving Decisions about Health, Wealth, and Happiness*. New Haven: Yale University Press/New York: Penguin.

Theiner, George, and T. O'Connor. 2010. "The Emergence of Group Cognition." In *Emergence in Science and Philosophy*, edited by A. Corradini and T. O'Connor, 78–117. New York: Routledge.

Thoreau, Henry David. 1993. "Life without Principle." In *Walden and Other Writings*. New York: Barnes & Noble Books.

Tilhonen, J., M.-R. Rautainen, H. M. Ollila, E. Repo-Tiihonen, M. Virkkunen, A. Palotie, O. Pietiläinen, K. Kristiansson, M. Jourkamaa, H. Lauerma, J. Saarela, S. Tyni, H. Vartiainen, J. Paananen, D. Goldman, and T. Paunio. 2015. "Genetic Background of Extreme Violent Behavior." *Molecular Psychiatry* 20: 786–92.

Torrey, E. Fuller. 1992. *Freudian Fraud: The Malignant Effect of Freud's Theory on American Thought and Culture*. Bethesda, MD: Lucas Books.

Turkle, Sherry. 1999. "Cyberspace and Identity." *Contemporary Sociology* 28(6): 643–48.

_____. 2012. *Alone Together: Why We Expect More from Technology and Less from Each Other*. New York: Basic Books.

Tuomela, Raimo. 2005. "We-Intentions Revisited." *Philosophical Studies* 125: 327–69.

Tulane, S., L. Skogran, and J. DeFrain. 2011. "Couples in Great Marriages Who Considered Divorcing." *Marriage and Family Review* 47: 289–310.

UNEP. 2015. "Issues in Sustainable Development." Accessed April 3, 2015. http://www.unep.org/training/programmes/Instructor%20Version/Part_2/Activities/Interest_Groups/Public_Awareness/index.html.

Utz, S., and C. J. Beukeboom. 2011. "The Role of Social Network Sites in Romantic Relationships: Effects on Jealousy and Relationship Happiness." *Journal of Computer-Mediated Communication* 16: 511–17.

Vaihinger, Hans. 1949. *The Philosophy of As If*. Translated by C. K. Ogden. London: Routledge & Kegan Paul.

Valenzuela, S., N. Park, and K. F. Kee. 2009. "Is There Social Capital in a Social Network Site? Facebook Use and College Students' Life Satisfaction, Trust, and Participation." *Journal of Computer-Mediated Communications* 14: 875–901.

Valéry, Paul. 1960. "La Conquête de l'Ubiquité." In *Œuvres, Tome II*, 1283–87. Paris: Gallimard.

van Kammen, Jessica. 2003. "Who Represents the Users? Critical Encounters between Women's Health Advocates and Scientists in Contraceptive R&D." In *How Users Matter: The Co-Construction of Users and Technology*, edited by N. Oudshoorn and T. Pinch. Cambridge, MA: MIT Press.

Verbeek, Peter-Paul. 2008. "Obstetric Ultrasound and the Technological Mediation of Morality: A Postphenomenological Analysis." *Human Studies* 31(1): 11–26.

_____. 2011a. *De grens van de mens.Over techniek, ethiek en de menselijke natuur.* [The boundary of the human being: On technology, ethics and human nature.] Rotterdam: Lemniscaat.

_____. 2011b. *Moralizing Technology: Understanding and Designing the Morality of Things*. Chicago: University of Chicago Press.

_____. 2012. "Expanding Mediation Theory." *Foundations of Science* 17(4): 391–95.

Vygotsky, Lev. 1997. "The Problem of the Development of Higher Mental Functions." Translated by Marie J. Hall. In *The Collected Works of L.S. Vygotsky: Volume 4*, edited by R. W. Rieber, 1–26. New York: Plenum Press.

Wajcman, Judy. 1991. *Feminism Confronts Technology*. University Park: Pennsylvania State University Press.

_____. 1995. "Feminist Theories of Technology." In *Handbook of Science and Technology Studies*, edited by J. Sheila et al., 189–204. Thousand Oaks: Sage Publications.

Want, Stephen C., Kristin Vickers, and Jennifer Amos. 2009. "The Influence of Television Programs on Appearance Satisfaction: Making and Mitigating Social Comparisons to 'Friends.'" *Sex Roles* 60: 642–55.

Ward, Adrian F. 2013. "Supernormal: How the Internet is Changing our Memories and our Minds." *Psychological Inquiry* 24: 341–48.

Warren, Samuel D., and Louis D. Brandeis. 1890. The Right to Privacy. *Harvard Law Review* 4(5): 193–220.

Watts, D. 2014. "Stop Complaining About the Facebook Study. It's a Golden Age of Research." *Guardian* (London). July 7.

Wegner, D. 1986. "Transactive Memory: A Contemporary Analysis of the Group Mind." In *Theories of Group Behavior*, edited by B. Mullen and G. Goethal, 185–208. New York: Springer-Verlag.

Weindling, P. J. 2004. *Nazi Medicine and the Nuremberg Trials: From Medical War-crimes to Informed Consent*. London: Palgrave Macmillan.

Wetmore, Daniel. 2005. "You Say 'Pre-Emptive,' I 'Precipitive . . .'" *The Joint Services Conference on Professional Ethics* (JSCOPE). Isme.tamu.edu/JSCOPE05/jscope05.html. Accessed 06/13/2015.

Whitty, M. T. 2005. "The Realness of Cyber Cheating: Men's and Women's Representations of Unfaithful Internet Relationships." *Social Science Computer Review* 23: 57–67.

Whetstine, Leslie M. 2015. "Ethical Challenges in Assisted Reproduction: The Place of Preimplantation Genetic Diagnosis in a Just Society." *Journal of Child Neurology* 30(5): 547–50.

Wiederhold, B. K. 2012. "As Parents Invade Facebook, Teens Tweet More." *Cyberpsychology, Behavior, and Social Networking* 15: 385.

Wiener, Norbert. 1948. *Cybernetics, Or, the Control and Communication in the Animal and the Machine*. London: MIT Press.

Wilson, Anne, and Michael Ross. 2003. "The Identity Function of Autobiographical Memory: Time is on our Side." *Memory* 11(2): 137–49.

Wilson, B., ed. 1970. *Rationality*. Oxford: Blackwell.

Wittes, Benjamin, and Jane Chong. 2014. "Our Cyborg Future: Law and Policy Implications." *The Brookings Institute*. http://www.brookings.edu/research/reports2/2014/09/cyborg-future-law-policy-implications.

Wittkower, D., Evan Selinger, and Lucinda Rush. 2014. "Public Philosophy of Technology: Motivations, Barriers, and Reforms." *Techné: Research in Philosophy and Technology* 17(2): 179–200.

Wyatt, Sally. 2008. "Technological Determinism is Dead: Long Live Technological Determinism." In *The Handbook of Science and Technology Studies*, edited by E. J. Hackett, O. Amsterdamska, M. Lynch, and J. Wajcman, 165–80. Cambridge, MA: MIT Press.

Young, K. S., E. Griffin-Shelley, A. Cooper, J. O'Mara, and J. Buchanan. 2000. "Online Infidelity: A New Dimension in Couple Relationships with Implications for Evaluation and Treatment." *Sexual Addiction and Compulsivity* 7: 59–74.

Zagzebski, Linda, ed. 2012. *Epistemic Authority: A Theory of Trust, Authority, and Autonomy in Belief.* Oxford: Oxford University Press.

Zahavi, D. 2001. "Beyond Empathy: Phenomenological Approaches to Intersubjectivity." In *Journal of Consciousness Studies* 8(5–7): 151–67.

_____. 2005. *Subjectivity and Selfhood: Investigating the First-Person Perspective.* Cambridge, MA: MIT Press.

Zea, Leopoldo. 1987. "Philosophy as an Instrument of Interamerican Understanding." *Social Epistemology: A Journal of Knowledge, Culture and Policy* 1(2): 123–30.

Zielinski, Siegfried. 2006. *Deep Time of the Media: Toward an Archaeology of Hearing and Seeing by Technical Means*. Cambridge, MA: MIT Press.

Index

About the Contributors

Steve Fuller is Auguste Comte professor of social epistemology in the Department of Sociology at the University of Warwick, United Kingdom. Originally trained in history and philosophy of science, Fuller is best known for his foundational work in the field of "social epistemology," which is the name of a quarterly journal that he founded in 1987 as well as the first of his more than twenty books. His most recent work has been concerned with the future of "humanity" as both a being and a concept.

Hans Radder is professor emeritus in philosophy of science and technology at the Department of Philosophy of VU University Amsterdam. He edited the volume *The Commodification of Academic Research: Science and the Modern University* (University of Pittsburgh Press, 2010), and is coeditor (with Alfred Nordmann and Gregor Schiemann) of *Science Transformed? Debating Claims of an Epochal Break* (University of Pittsburgh Press, 2011). One of his interests in philosophy of technology is patenting (see "Exploring Philosophical Issues in the Patenting of Scientific and Technological Inventions," *Philosophy & Technology*, 2013).

Daniel J. Brunson is a lecturer in the Department of Philosophy and Religious Studies at Morgan State University in Baltimore. Daniel's current research continues to engage the insights of classically informed pragmatism with contemporary social epistemology, ethics, and science and technology. In particular, he is interested in the intersection of knowledge and morality under the topic of "risk," such as his article "Insuring the Community against Loss: Roycean Reflections on the Tasks of Interpretation."

Jamie Carlin Watson is assistant professor of philosophy at Broward College, Ft. Lauderdale, Florida. His primary research is in the social epistemology of expertise and epistemic authority. He has articles in *Journal of Applied Philosophy*, *Episteme*, and *Contemporary Debates in Bioethics* (Blackwell, 2013). He has coauthored *What's Good on TV? Understanding Ethics through Television* (Wiley-Blackwell, 2011) (with Robert Arp) and *The Critical Thinking Toolkit* (Wiley-Blackwell, 2015) (with Peter Fosl and Galen Foresman), with the monograph *Fear, Pity, Guilt: Winning Votes by Abusing Reason* (Lexington Books, forthcoming).

Patrick J. Reider, PhD, is an assistant professor at the University of Pittsburgh Greensburg. His primary interests are in social epistemology and German idealism. Currently, he is an editor and contributing author for two books: *Wilfrid Sellars, Idealism and Realism: Understanding Psychological Nominalism* (Bloomsbury Publishing) and *Social Epistemology and Epistemic Agency: Decentralizing the Epistemic Agent* (Rowman and Littlefield International). He is the special issue editor for *Social Epistemology Review and Reply Collective*, the online platform for the journal *Social Epistemology: A Journal of Knowledge, Culture and Policy*, by Taylor & Francis.

Elize de Mul holds Master's degrees in new media studies and philosophy. Currently, she is a PhD candidate at the eLaw—Centre for Law in the Information Society—of Leiden University (NL) conducting research on technology, identity, and privacy. She is also lecturer of philosophy at ArtEZ, Institute of the Arts. Among her publications are "Telling your Life Story," in Lorenzo von Matterhorn (ed.), *How I Met Your Mother and Philosophy* (2013), and a philosophical essay on the plastic bag, *Dancing with a Plastic Bag* (2013, in Dutch).

Stephen M. Bourque is a PhD candidate in philosophy at Temple University. His articles have appeared in such journals as *ViraVerita*, *Rivista di esterica*, and *Socrates*. He is interested in Marxist thought and critical media studies.

William Davis, MA, MS, is a PhD candidate in science and technology studies at Virginia Tech. His particular focus, philosophy of technology, gives rise to his notion of an "un-disciplined" philosophy of technology that transcends disciplines and engages broad audiences. His dissertation argues that the ethical, legal, political, and social implications of emerging technologies, and their concomitant values, must become part of general public discussions if we wish to exercise influence on the development and deployment of the technologies that will help shape our future lives and societies.

Charles Bambach is professor of philosophy at the University of Texas at Dallas. His latest book *Thinking the Poetic Measure of Justice: Heidegger, Hölderlin, Celan* appeared in 2013 in the SUNY Series in Contemporary Continental Philosophy. Other books include *Heidegger's Roots: Nietzsche,*

National Socialism, and the Greeks (2003) and *Heidegger, Dilthey, and the Crisis of Historicism* (1995), both from Cornell University Press. He has also written a variety of articles on hermeneutics, phenomenology, ethics, Nietzsche, and the history of German and ancient Greek philosophy.

Joris Vlieghe studied philosophy and art history, and obtained his PhD in educational sciences on an investigation into the public and educational meaning of corporeality. He teaches philosophical foundations of education, and ethics and education at Moray House School of Education (Edinburgh University). His research focuses on how the growing presence of digital technologies alters existing school practices, and the relation between this evolution and new forms of subjectivity, including the shift from book-culture to screen-culture, and its consequences regarding basic educational concepts such as literacy, creativity, and transformation.

Arthur Kok is a postdoctoral researcher and lecturer at Tilburg University. He teaches social philosophy and history of modern philosophy, specializing in modern and contemporary philosophy, in particular the philosophy of Immanuel Kant and German Idealism with a focus on the relationship between philosophy, religion, and science. His recent publications include "Metaphysics of Recognition," "Labor as Recognition: An Hegelian Approach to Meta-economics," and "Contemporary Social Contract Theory and Hegel's Master/Bondsman-relation."

Danielle Guizzo is a PhD candidate at the Graduate Program of Public Policy, Federal University of Parana, and works in the field of philosophy of social sciences, exploring such topics as economic methodology, philosophy of economics, and gender studies. She has recently published in the area of economic methodology, science, technology, and society, and history of economic thought.

Mindaugas Briedis is full professor at Mykolas Romeris University, specifically the Institute of Humanities and department of philosophy located in Vilnius, Lithuania. He is author of more than thirty articles, Fulbright nominee (2013), and keynote speaker at the Baltic Congress of Radiology (2012). His research interests include classical phenomenology, transcendental phenomenology and media, and philosophy of medicine.

Chad Engelland, PhD, is assistant professor of philosophy at the University of Dallas. He is the author of *Ostension: Word Learning and the Embodied Mind* (MIT Press, 2014) and *The Way of Philosophy* (Cascade Books, 2016). His articles have appeared in such journals as *Continental Philosophy Review*, *Journal of the British Society for Phenomenology*, and *Journal of the American Philosophical Association*.

Levi Checketts is a PhD candidate at the Graduate Theological Union in Berkeley, California, where his work centers on the intersection of Catholic moral theology, existential phenomenology, and technological ethics. He works as an adjunct professor at Holy Names University in the philosophy department.

Richard Charles Strong is a PhD candidate in the philosophy department at Villanova University. His dissertation project examines and refines the related concepts of habitus, habit, and second nature from complementary perspectives in critical sociology, embodied phenomenology, technology studies, and philosophy of mind. He is author of "Habit and the Extended Mind: Fleshing Out The Extended Mind Theory with Merleau-Pontian Phenomenology" in *Phenomenology and Mind* and "Habit, Distraction, Absorption: Reconsidering Walter Benjamin and the Relation of Architecture to Film" in *The Missed Encounter of Radical Philosophy with Architecture.*

Chris Drain is a PhD candidate in the Department of Philosophy at Villanova University. He works at the intersection of philosophy of technology, philosophy of mind, and cognitive science. His dissertation reflects on the relevance of the Vygotsky circle for contemporary discourses surrounding agency and signification in the philosophy of mind and cognitive anthropology. His broader academic interests include cognitive ethology, social ontology, and the history of modern philosophy.

Frank Scalambrino is Senior Leturer at the University of Akron, Ohio's Polytechnic University. He has taught graduate courses in both philosophy and psychology in Illinois and Texas. Author of *Full Throttle Heart: Nietzsche Beyond Either/Or* (The Eleusinian Press, 2015), his work has appeared in *The Review of Metaphysics*, *Philosophical Psychology*, *Phenomenology and Mind*, *Topos*, *Reason Papers*, *Philosophy in Review*, *Social Epistemology Review and Reply Collective*, and the *Internet Encyclopedia of Philosophy*. An advocate for "existentialism" and "public philosophy," he is critical of cybernetics, especially its application in psychology.

www.ingramcontent.com/pod-product-compliance
Lightning Source LLC
LaVergne TN
LVHW101446200926
848483LV00001B/66
9781783485321